Leland Ossian Howard

Principal Household Insects of the United States

Leland Ossian Howard

Principal Household Insects of the United States

ISBN/EAN: 9783337404581

Printed in Europe, USA, Canada, Australia, Japan

Cover: Foto ©berggeist007 / pixelio.de

More available books at **www.hansebooks.com**

BULLETIN No. 4.—NEW SERIES.

U. S. DEPARTMENT OF AGRICULTURE.
DIVISION OF ENTOMOLOGY.

THE PRINCIPAL

HOUSEHOLD INSECTS

OF THE

UNITED STATES.

BY

L. O. HOWARD AND C. L. MARLATT.

WITH A CHAPTER ON

INSECTS AFFECTING DRY VEGETABLE FOODS,

BY

F. H. CHITTENDEN.

LETTER OF TRANSMITTAL.

U. S. DEPARTMENT OF AGRICULTURE,
DIVISION OF ENTOMOLOGY,
Washington, D. C., July 7, 1896.

SIR: I have the honor to submit for publication the accompanying account of the principal household insects of the United States.

Respectfully,

L. O. HOWARD,
Entomologist.

Hon. J. STERLING MORTON,
Secretary of Agriculture.

2

CONTENTS.

ILLUSTRATIONS.

INTRODUCTION.

On an average, from 500 to 600 letters of inquiry are received at this office each month. A very considerable number of these inquiries relate to insects which are found in houses and which either annoy the occupants by their direct attacks or are injurious to household goods and provisions. The available literature on this class of insects is not extensive. Prof. C. H. Fernald, of the Massachusetts Agricultural Experiment Station, published a short bulletin on the general subject some three years ago, but only a few of the most prominent insects of this class were treated. Other American articles are scattered in various publications, in the reports of the State entomologists and bulletins of the entomologists of the State agricultural experiment stations, and in the entomological and other scientific journals. A small volume was published in England in 1893, which bears the title of Our Household Insects, by Mr. Edward A. Butler, a competent entomologist, who has brought together a mass of interesting facts. This little volume, however, treats of English insects only. There is abundant room, then, for the present publication. Much that is presented herewith is based upon original observations in the office, and all accessible publications upon the species treated have been consulted. As will be observed from the title-page, the preparation of the bulletin has been the joint work of the writer and of Messrs. Marlatt and Chittenden. Mr. Chittenden's work has been confined to a concluding chapter on the subject of the species that affect dry vegetable foods, a labor for which he is particularly well fitted by reason of his long study of these species. There has been no systematic division in the work of the main portion of the bulletin between the writer and Mr. Marlatt. Each of us has chosen the topics in which he felt especially interested. It results that longer or shorter articles by one or the other are arranged according to the proper position of the topic in the scheme as a whole and are not brought together under the respective authors. The authorship of the individual articles, however, may be readily accredited by the fact that not only is it displayed in the table of contents, but by the further fact that the contributions are initialed in every case.

The very curious but not unexpected condition has been shown in the preparation of this bulletin that of some of our commonest household insects the life history is not known with any degree of exactness.

Of such common species as the household centipede (*Scutigera forceps*) and the "silver fish" or "slicker" (*Lepisma* spp.) careful studies yet remain to be made, and it is hoped that one of the incidental benefits which will result from the publication of this bulletin will be this indication of topics of desired investigation to students. The illustrations have all been made by Miss Sullivan, with the exception of those of the cheese skipper and ham beetles and the house centipede, which have been prepared by Mr. Otto Heidemann. All drawings have been made under the supervision of the author of the section in which they appear.

L. O. H.

THE PRINCIPAL HOUSEHOLD INSECTS

OF THE

UNITED STATES.

CHAPTER I.

MOSQUITOES AND FLEAS.

By L. O. Howard.

MOSQUITOES.

(*Culicidæ* spp.)

Although mosquitoes are out-of-door insects, they may be considered appropriately under the head of household pests, for the reason that they enter houses, to the torment of the inhabitants, all through the summer months, and many of them pass the winter in cellars. In fact, it is probably safe to say that no distinctive household pest causes as much annoyance as the mosquito.

We are accustomed to think and speak of the mosquito as if there were but one species; yet, to our knowledge, there are no less than eight species, for example, which are more or less common in the District of Columbia, and the writer has noticed at New Orleans, La., certainly four different species at the same season of the year, while at Christmas time a fifth species, smaller than the others, causes considerable trouble in the houses of that city. In Trinidad Mr. Urich states that he has observed at least ten different species from the island of St. Vincent. In his Catalogue of the Diptera of North America Baron Osten Sacken records twenty-one from North America, and it is perhaps safe to say that not half of the species are described. In the collection of the United States National Museum there are twenty distinct species, all of which have been authentically determined by Mr. Coquillett.

The common species at Washington in the months of May and June is *Culex pungens* Wied. I say the *common* species, but do not wish to be understood as saying that mosquitoes are common in Washington at that time of the year. As a matter of fact, the city is singularly free from this little pest, and this is largely due to the reclamation of the marshes of the Potomac River, which in war times and for a number of years afterwards caused the inhabitants of this city to suffer severely from this insect. As late as 1875, it is said, it was almost impossible to spend any of the night hours near the marshes without smudges. Later in the season other species become abundant.

The writer, in the course of certain observations, has carried *C. pungens* through approximately two generations in the early part of the season. It is strange that recent and definite observations upon accu-

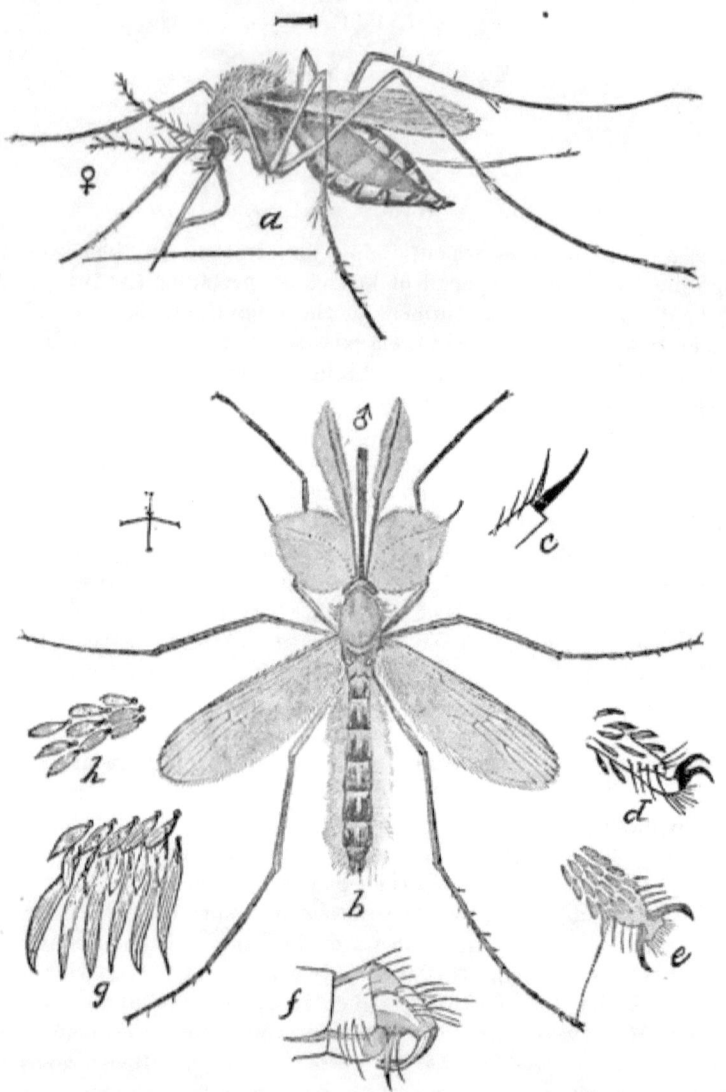

Fig. 1.—*Culex pungens: a*, female, from side; *b*, male, from above; *c*, front tarsus of same; *d*, middle tarsus; *e*, hind tarsus; *f*, genitalia of same; *g*, scales from hind border of wing; *h*, scales from disk of wing—enlarged (original).

rately determined species of many of our commonest insects have not been published. This is mainly due to the fact that most entomologists have a way of saving time by following the observations of older writers.

This is all well enough where the species and the conditions are identical, but when, as is the case with such an insect as that under observation, the principal observations were made upon a different, though congeneric, species, and in another part of the globe, where climatic and other conditions differ, the custom is unfortunate. There is not, in any of our published works, a thoroughly satisfactory figure of a well-determined species of mosquito, or of its earlier stages. The statements quoted in the text-books and manuals date back, in general, to the time of Réaumur, one hundred and fifty years ago. These observations were made in the month of May, upon a species (*Culex pipiens*) which does not occur in North America, and in the one locality of Paris, France. The notes made upon *C. pungens* at Washington possess, therefore, some scientific importance.

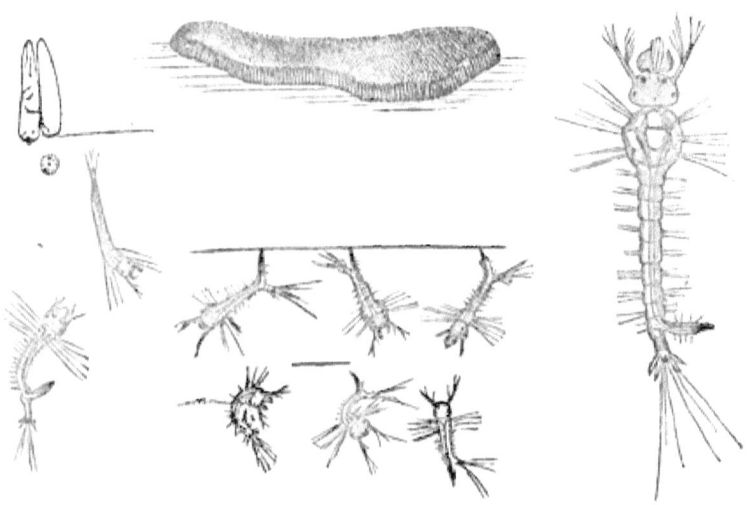

FIG. 2.—*Culex pungens*: Egg-mass above in center; young larva, greatly enlarged, at right; young larva, not so much enlarged, below; enlarged eggs above at left (original).

The operation of egg-laying was not observed, but it probably takes place in the very early morning hours. The eggs are laid in the usual boat-shaped mass, just as those of *C. pipiens*, as described by Réaumur. We say boat-shaped mass, because that is the ordinary expression. As a matter of fact, however, the egg masses are of all sorts of shapes. The most common one is the pointed ellipse, convex below and concave above, all the eggs perpendicular, in six to thirteen longitudinal rows, with from 3 or 4 to 40 eggs in a row. The number of eggs in each batch varies from 200 to 400. As seen from above, the egg-mass is gray brown; from below, silvery white, the latter appearance being due to the air film. It seems impossible to wet these egg masses. They may be pushed under water, but bob up, apparently as dry as ever. The egg mass separates rather regularly and the eggs are not stuck together very firmly. After they have hatched the mass will disintegrate in a few days, even in perfectly still water.

The individual eggs are 0.7 mm. in length and 0.16 mm. in diameter at the base. They are slender, broader and blunt at bottom, slenderer and somewhat pointed at tip. The tip is always dark grayish brown in color, while the rest of the egg is dirty white. Repeated observations show that the eggs hatch, under advantageous conditions, certainly as soon as sixteen hours. Water buckets containing no egg masses, placed out at night, were found to contain egg masses at 8 o'clock in the morning, which, as above stated, were probably laid in the early morning, before daylight. These eggs, the third week in May, began to hatch quite regularly at 2 o'clock in the afternoon of the same day on warm days. In cooler weather they sometimes remained unhatched until the second day. If we apply the evidence of European observers to this species, the period of the egg state may be under twelve hours; but there is a possibility that they are laid earlier in the night, which accounts for the fact that sixteen hours is the shortest period which we can definitely mention.

The larvæ issue from the underside of the egg masses, and are extremely active at birth. When first observed it is easy to fall into an error regarding the length of time which they can remain under water, or rather without coming to the surface to breathe, since, in striving to come to the surface for air, many of them will strike the underside of the egg mass and remain there for many minutes. It is altogether likely, however, that they get air at this point through the eggs or through the air film by which the egg mass is surrounded, and that they are as readily drowned by continuous immersion as are the older ones, as will be shown later.

One of the first peculiarities which strikes one on observing these newly hatched larvæ under the lens is that the tufts of filaments which are conspicuous at the mouth are in absolutely constant vibration. This peculiarity, and the wriggling of the larvæ through the water, and their great activity, render them interesting objects of study. In general, the larvæ, passing through apparently three different stages, reach maturity and transform to pupæ in a minimum of seven days. When nearly full grown their movements were studied with more care, as they were easier to observe than when newly hatched. At this time the larva remains near the surface of the water, with its respiratory siphon at the exact surface and its mouth filaments in constant vibration, directing food into the mouth cavity. Occasionally the larva descends to the bottom, but, though repeatedly timed, a healthy individual was never seen to remain voluntarily below the surface more than a minute. In ascending it comes up with an effort, with a series of jerks and wrigglings with its tail. It descends without effort, but ascends with difficulty; in other words, its specific gravity seems to be greater than that of the water. As soon, however, as the respiratory siphon reaches the surface, fresh air flows into its trachea, and the physical properties of the so-called surface film of the water assist it in maintaining its position.

The account by Miall, in his recently published Natural History of Aquatic Insects, is misleading, for the reason that he assumes that the end of the body, with its four (or, as he has it, five) leaf-like expansions, is the breathing organ. As a matter of fact, as is plainly shown by fig. 2, this end of the body does not reach the surface, and it is the tip of the respiratory siphon only which is extended to the air. This respiratory tube takes its origin from the tip of the eighth abdominal segment, and the very large trachea can be seen extending to its extremity, where they have a double orifice. The ninth segment of the abdomen is armed at the tip with four flaps and six hairs, as shown in fig. 4. These flaps are gill-like in appearance, though they are probably simply locomotory in function. With so remarkably developed an

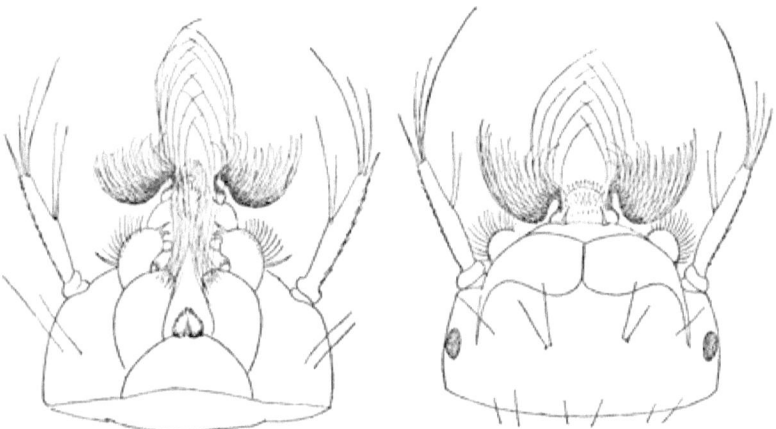

Fig. 3.—*Culex pungens*: Head of larva from below at left, same from above at right—greatly enlarged (original).

apparatus for direct air breathing there is no necessity for gill structures. Raschke[1] and Hurst[2] consider that the larva breathes both by the anus and by these gill flaps, as well as by the large trachea which open at the tip of the respiratory tube. Raschke considers that these trachea are so unnecessarily large that they possess a hydrostatic function. The writer is inclined to believe that the gill flaps may be functional as branchial structures in the young larva, but that they largely lose this office in later life.

After seven or eight days, at a minimum, as just stated, the larva transforms to pupa. The pupa, as has been repeatedly pointed out with other species, differs most pronouncedly from the larva in the great swelling of the thoracic segments. In this stage the insect is

---

[1] Raschke, Die Larve von *Culex nemorosus*, Berlin, 1887.
[2] Hurst, The Pupal Stage of Culex, Manchester, 1890.

lighter than water. It remains motionless at the surface, and when disturbed does not sink without effort, as does the larva, but is only able to descend by a violent muscular action. It wriggles and swims as actively as does the larva, and soon reaches the bottom of the jar or breeding place. As soon as it ceases to exert itself, however, it floats gradually up to the surface of the water again. The fact, however, that the larva, after it is once below the surface of the water, sinks rather than rises, accounts for the death of many individuals. If they become sick or weak, or for any reason are unable to exert sufficient muscular force to wriggle to the surface at frequent intervals, they will actually drown, and the writer has seen many of them die in this way. It seems almost like a contradiction in terms to speak of an aquatic insect drowning, but this is a frequent cause of mortality among wrigglers. This fact also explains the efficacy of the remedial treatment which causes the surface of the water to become covered with a film of oil of any kind. Aside from the actual insecticide effect of the oil, the larvæ drown from not being able to reach the air. The structure of the pupa differs in no material respect from that of corresponding stages of European species, as so admirably figured and described by the older writers, notably Réaumur and Swammerdam,[1] and needs no description in view of the care with which the figures accompanying this article have been drawn. The air tubes no longer open at the anal end of the body, but through two trumpet-shaped sclerites on the thorax, from which it results that the pupa remains upright at the surface, instead of with the head downward. There is a very apparent object in this reversal of the position of the body, since the adult insect issues from the thorax and needs the floating skin to support itself while its wings are expanding.

In general, the adult insects issue from the pupæ that are two days old. This gives what is probably the minimum generation for this species as ten days, namely, sixteen to twenty-four hours for the egg, seven days for the larva, and two days for the pupa. The individuals emerging on the first day were invariably males. On the second day the great majority were males, but there were also a few females. The preponderance of males continued to hold for three days; later the females were in the majority. In confinement the males died quickly; several lived for four days, but none for more than that period. The females, however, lived for a much longer time. Some were kept alive without food, in a confined space of not more than 4 inches deep by 6 across, for three weeks. But one egg mass was deposited in confinement. This was deposited on the morning of June 30 by a female which issued from the pupa June 27. No further observations were made upon the time elapsing between the emergence of the female and the laying of the eggs, but in no case, probably, does it exceed a few days.

---

[1] Even Bonanni, in 1691, gave very fair figures of the larva and pupa of a European species. Micrographia Curiosa, Rome, MDCXCI, Pars. II, Tab. I.

The length of time which elapses for a generation, which we have just mentioned, is almost indefinitely enlarged if the weather be cool. As a matter of fact, a long spell of cool weather followed the issuing of the adults just mentioned. Larvæ were watched for twenty days, during which time they did not reach full growth.

The extreme shortness of this June generation is significant. It accounts for the fact that swarms of mosquitoes may develop upon occasion in surface pools of rain water, which may dry up entirely in

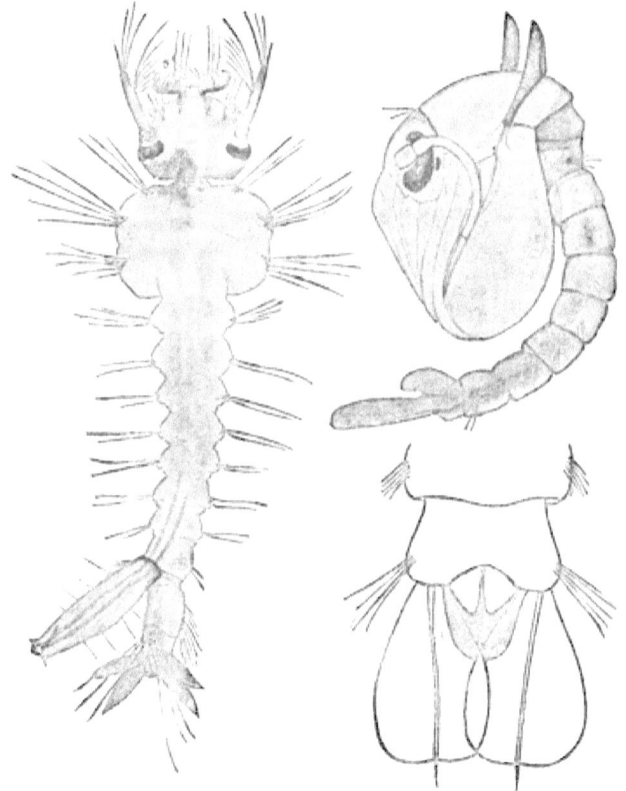

Fig. 4.—*Culex pungens:* Full-grown larva at left; pupa at right above, its anal segment below—all greatly enlarged (original).

the course of two weeks, or in a chance bucket of water left undisturbed for that length of time. Further, the shortness of this generation was, while not unexpected, not at all in accordance with any published statements as to the length of life of any immature mosquito of any species. But these published statements, as previously shown, were nearly all based upon observations made in a colder climate and in the month of May.

On August 1 Mr. F. C. Pratt, an assistant in the division of entomology, brought in from Lakeland, Md., a small place 9 miles from Washington, specimens of a large and very ferocious mosquito, which Mr. Coquillett determined as *Anopheles quadrimaculatus* Say, a species which had previously been observed at Washington in August. This mosquito was very abundant at Lakeland at the time, and its eggs were obtained, but rearing operations were interrupted by absence from Washington. At the same time the commonest of the mosquitoes at Washington was found to be *Culex consobrinus*. This latter species was one which was studied by the writer in 1892 in the Catskill Mountains, near Tannersville, Greene County, N. Y. This species in Washington became, during August, more abundant than *C. pungens*. October 25, however, the writer found both species in his house, which they had evidently entered for hibernation. In 1893 several specimens of *pungens* were taken in the month of January in the cellar of his house in Georgetown. This hibernation in cellars as well as in outhouses is very common, although it is not frequently referred to. Specimens of *C. consobrinus* were received in November, 1894, from J. M. Wade, of Boston, with the statement that they were abundant in his cellar in that city. The cellar was very cold, although in one corner there was a tin furnace pipe. The mosquitoes avoided the warm corner, and were always thickest in the cold parts of the cellar. So abundant were they that if a lamp were held up the inside of the chimney would soon be covered half an inch thick with their bodies.

The degree of cold seems to make no difference with this successful hibernation. Arctic explorers have long since recorded the abundance of mosquitoes in the extreme north. In the narrative of C. F. Hall's second arctic expedition the statement is made that mosquitoes appeared on the 7th of July, 1869, in extraordinary abundance. Dr. E. Sterling, of Cleveland, Ohio, has sent us an account of the appearance of mosquitoes by thousands in March, 1844, when he was on a snowshoe trip from Mackinaw to Sault Ste. Marie. Their extraordinary numbers at this season of the year is remarkable, indicating a most plentiful hibernation. Mr. H. Stewart, of North Carolina, has written us of a similar experience on the north shore of Lake Superior in 1866. On warm days in March, when the snow was several feet deep and the ice on the lake 5 feet in thickness, mosquitoes appeared in swarms, "literally blackening the banks of snow in the sheltered places." The Indians told Mr. Stewart that the mosquitoes lived through the winter, and that the old ones were the most annoying to them. May 9, 1896, Mr. Lugger sent the writer from St. Anthony Park, Minn., specimens of *C. consobrinus*, stating that it came in a genuine swarm in April, with a heavy snowstorm, at a time when all of the lakes were covered with ice—"Minnesota's most certain crop."

It is a well-known fact that the adult male mosquito does not necessarily take nourishment, and that the adult female does not necessarily rely upon the blood of warm-blooded animals. They are plant feeders and have also been recorded as feeding upon insects. Dr. Hagen mentions taking a species in the Northwest feeding upon the chrysalis of a butterfly, while scattered through the seven volumes of Insect Life are a number of records of observations of a vegetarian habit, one writer stating that he has seen them with their beaks inserted in boiled potatoes on the table, and another that he has seen watermelon rinds with many mosquitoes settled upon them and busily engaged in sucking the juices. Mosquitoes undoubtedly feed normally on the juices of plants, and not one in a million ever gets an opportunity to taste the blood of a warm-blooded animal. When we think of the enormous tracts of marsh land into which warm-blooded animals never penetrate, and in which mosquitoes are breeding in countless numbers, the truth of this statement becomes apparent. The males have been observed sipping at drops of water, and one instance of a fondness for molasses has been recorded. Mr. E. A. Schwarz has observed one drinking beer.

The literature of popular entomology is full of instances of the enormous numbers in which mosquitoes occasionally occur, but a new instance may not be out of place here. Mr. Schwarz tells the writer that he has never seen, even in New Jersey, mosquitoes to compare in numbers with those at Corpus Christi, Tex. When the wind blows from any other direction than south, he says, hundreds of thousands of millions of mosquitoes blow in upon the town. Great herds of hundreds of horses run before the mosquitoes in order to get to the water. With a change of wind, however, the mosquitoes blow away..

REMEDIES AGAINST MOSQUITOES.

Of the remedies in use in houses the burning of pyrethrum powder and the catching of the mosquitoes on the walls with kerosene in cups, as described in Insect Life (Vol. V, p. 143), are probably the best, next to a thorough screening and mosquito bars about the bed. It may be of interest to mention incidentally a remedy in use among the Chinese, as recorded in Robert Fortune's "Residence Among the Chinese: Scenes and Adventures Among the Chinese in 1853-1856" (London, 1857). Long-necked bags of paper, half an inch in diameter and 2 feet long, are filled with the following substances: Either pine or juniper sawdust, mixed with a small quantity of "nu-wang" and 1 ounce of arsenic. These substances are well mixed and run into the bags in a dry state; each bag is coiled like a snake and wrapped and tied with thread. The outer end is lighted and the coil laid on a board. Two coils are sufficient for an ordinary-sized room, and 100 coils sell for 6 cents. Mr. Mun Yen Chung, of the Chinese legation, has been good enough to inform the writer that by "nu-wang" Mr. Fortune probably meant liu-wang (brimstone).

Altogether the most satisfactory ways of fighting mosquitoes are those which result in the destruction of the larvæ or the abolition of their breeding places. In not every locality are these measures feasible, but in many places there is absolutely no necessity for the mosquito annoyance. The three main preventive measures are the draining of breeding places, the introduction of small fish into fishless breeding places, and the treatment of such pools with kerosene. These are three alternatives, any one of which will be efficacious, and any one of which may be used where there are reasons against the trial of the others.

In 1892 the writer published the first account of extensive out-of-doors experiments to determine the actual effect upon the mosquitoes of a thin layer of kerosene upon the surface of water in breeding pools and the relative amount to be used. He showed the quantity of kerosene necessary for a given water surface, and demonstrated further that not only are the larvæ and pupæ thereby destroyed almost immediately, but that the female mosquitoes are not deterred from attempting to oviposit upon the surface of the water, and that they are thus destroyed in large numbers *before their eggs are laid.* He also showed approximately the length of time for which one such treatment would remain operative. No originality was claimed for the suggestion, but only for the more or less exact experimentation. The writer himself, as early as 1867, had found that kerosene would kill mosquito larvæ, and the same knowledge was probably put in practice, although without publicity, in other parts of the country. In fact, Mr. H. E. Weed states (Insect Life, Vol. VII, p. 212) that in the French quarter of New Orleans it has been a common practice for many years to place kerosene in the water tanks to lessen the numbers of mosquitoes in a given locality, although he knew nothing that had been written to show that such was the case, and he says: "In this age of advancement we can no longer go by hearsay evidence." Suggestions as to the use of kerosene, and even experiments on a water surface 10 inches square, showing that the larvæ could be killed by kerosene, were recorded by Mrs. C. B. Aaron in her Lamborn prize essay and published in the work entitled "Dragon Flies *versus* Mosquitoes" (D. Appleton & Co., 1890). Mr. W. Beutenmüller also in the same work made the same suggestion.

The quantity of kerosene to be practically used, as shown by the writer's experiments, is approximately 1 ounce to 15 square feet of water surface, and ordinarily the application need not be renewed for one month. Since 1892 several demonstrations, on both a large and a small scale, have been made. Two localities were rid of the mosquito plague under the supervision of the writer by the use of kerosene alone. Mr. Weed, in the article above mentioned, states that he rid the college campus of the Mississippi Agricultural College of mosquitoes by the treatment with kerosene of eleven large water tanks. Dr. John B. Smith has recorded, though without details, success with this remedy in two cases on Long Island (Insect Life, Vol. VI, p. 91). Prof. J. H. Comstock tells the writer that a similar series of experiments, with perfectly satisfactory results, was carried out by Mr. Vernon L. Kellogg on the campus of Stanford University, at Palo Alto, Cal. In this case post holes filled with surface water were treated, with the result that the mosquito plague was almost immediately alleviated.

Additional experiments on a somewhat larger scale have been made by Rev. John D. Long at Oak Island Beach, Long Island Sound, and by Mr. W. R. Hopson, near Bridgeport, Conn., also on the shores of Long Island Sound, the experiments in both cases indicating the efficacy of the remedy when applied intelligently. I have not been able to learn the details of Mr. Hopson's operations, but am told that they included extensive draining as well as the use of kerosene.

It is not, however, the great sea marshes along the coast, where mosquitoes breed in countless numbers, which we can expect to treat by this method, but the inland places, where the mosquito supply is derived from comparatively small swamps and circumscribed pools. In most localities people endure the torment or direct their remedies against the adult insect only, without the slightest attempt to investigate the source of the supply, when the very first step should be the undertaking of such an investigation. In "Gleanings in Bee Culture" (October 1, 1895) we notice the statement in the California column that in some California towns the pit or vault behind water-closets is subject to flushing with water during the irrigation of the land near by. A period of several weeks elapses before more water is turned in, and in the meantime the water becomes stagnant and the breeding place of millions of mosquitoes. Then, as the correspondent says, "people go around wondering where all the mosquitoes come from, put up screens, burn buhach, and make a great fuss." Nothing could be easier than to pour an ounce of kerosene into each of these pits, and all danger from mosquitoes will have passed.

In many houses in Baltimore, Md., the sewage drains first into wells or sinks in the backyard, and thence in some cases into sewers, and in other cases is pumped out periodically. These wells invariably have open privies built over them, and the mosquitoes, which breed in the stagnant contents of the sinks, have free egress into the open air back of the houses. Hence parts of Baltimore much further removed from either running or stagnant water than certain parts of Washington, where no mosquitoes are found, are terribly mosquito ridden, and sleep without mosquito bars is, from May to December, almost impossible. Specimens of *Culex pungens* captured November 5 in such a privy as described have been brought to the writer from Baltimore by one of his assistants, Mr. R. M. Reese.

Kerosene has been tried by Mr. Reese in one case in Baltimore, and two treatments of a privy made about May 1 and June 1, respectively, seemed to diminish the numbers of the pest in that particular house; but without concerted action of all the householders in a given block (all the houses, be it remembered, being exactly alike in the method of sewage disposal) no great amount of good could be accomplished. With such concerted action, however, there seems to be no reason why the mosquito plague could not be greatly diminished in many, if not most, parts of Baltimore at a very small expense. Usually one well serves two houses, the privies being built in pairs, so that one treatment would suffice for two dwellings.

On ponds of any size the quickest and most perfect method of forming a film of kerosene will be to spray the oil over the surface of the water.

The remedy which depends upon draining breeding places needs no extended discussion. Naturally the draining off of the water of pools will prevent mosquitoes from breeding there, and the possibility of such draining and the means by which it may be done will vary with each individual case. The writer is informed that an elaborate bit of work which has been done at Virginia Beach bears on this method. Behind the hotels at this place, the hotels themselves fronting upon the beach, was a large fresh-water lake, which, with its adjoining swamps, was a source of mosquito supply, and it was further feared that it made the neighborhood malarious. Two canals were cut from the lake to the ocean, and by means of machinery the water of the lake was changed from a body of fresh to a body of salt water. Water that is somewhat brackish will support mosquitoes, but water which is purely salt will destroy them.

The introduction of fish into fishless breeding places is another matter. It may be undesirable to treat certain breeding places with kerosene, as, for instance, water which is intended for drinking, although this has been done without harm in tanks where, as is customary, the drinking supply is drawn from the bottom of the tank. An interesting case noted in Insect Life (Vol. IV, p. 223), in which a pair of carp was placed in each of several tanks, in the Riviera, is a case in point. The value of most small fishes for the purpose of destroying mosquito larvæ was well indicated by an experience described to us by Mr. C. H. Russell, of Bridgeport, Conn. In this case a very high tide broke away a dike and flooded the salt meadows of Stratford, a small town a few miles from Bridgeport. The receding tide left two small lakes, nearly side by side and of the same size. In one lake the tide left a dozen or more small fishes, while the other was fishless. An examination by Mr. Russell in the summer of 1891 showed that while the fishless lake contained tens of thousands of mosquito larvæ, that containing the fish had no larvæ.

The use of carp for this purpose has been mentioned in the preceding paragraph, but most small fish will answer as well. The writer knows of none that will be better than either of the common little sticklebacks (*Gasterosteus aculeatus* or *Pygosteus pungitius*). They are small, but very active and very voracious. Mr. F. W. Urich, of Trinidad, has written us that there is a little cyprinoid common in that island which answers admirably for this purpose. This fish has not been specifically determined, but we hope to make an effort to introduce it into our Southern States, if it proves to be new to our fauna. At Bee-ville. Tex., a little fish is used for this purpose which is there called a perch, although we have not been able to find out just what the species is. They soon eat up the mosquito larvæ, however, and in order to keep them alive the people adopt an ingenious fly trap, which they keep in their houses and in which about a quart of flies a day is caught. These flies are then fed to the fish. This makes a little circle which strikes us as particularly ingenious and pleasant. The fly traps catch the flies and rid the house of that pest. The flies are fed to the fish in the water tanks and keep them alive in order that they may feed on the mosquito larvæ, thus keeping the houses free of mosquitoes.

Where kerosene is considered objectionable, and where fish can not be readily obtained, there is another course left open. It is the con-stant artificial agitation of the water, since mosquitoes will oviposit only in still water. At San Diego, Tex., in the summer there are no streams for many miles, but plenty of mosquitoes breed in the water tanks. Some enterprising individuals keep their tanks free by putting in a little wheel, which is turned by the windmill, and keeps the water almost constantly agitated.

THE MOSQUITOES OF THE COUNTRY AT LARGE.

In the introductory paragraph the writer has indicated that we have numerous species among the mosquitoes of the United States and that several different species may occur in the same locality. It happens, however, that no definite knowledge exists, even among entomologists as to the exact species which may be found in any given locality. The desirability of a careful study of our mosquitoes is therefore apparent. As a preliminary step, the writer borrowed all of the mosquitoes from the collections of Prof. Lawrence Bruner, of the University of Nebraska, Lincoln, Nebr.; Prof. J. H. Comstock, of Cornell University, Ithaca, N. Y.; Prof. H. Garman, of the agricultural experiment station at Lexington, Ky.; Prof. C. P. Gillette, agricultural experiment station, Fort Collins, Colo.; Prof. C. W. Johnson, Wagner Free Institute, Phil-adelphia, Pa.; Prof. Otto Lugger, agricultural experiment station, St. Anthony Park, Minn.; Dr. W. A. Nason, Algonquin, Ill., and Mr. Th. Pergande, Washington, D. C. The material thus received, together with the collection of Culicidæ of the department of insects in the National Museum, was placed in the hands of Mr. D. W. Coquillett for specific study.

The results of this study were interesting. Mr. Coquillett had under his hands mosquitoes from nearly all portions of the United States. He found that the material represented twenty different species, of five genera, and was able to make out some important synonymical facts. In the distribution of certain species the results were unexpected. It was found that some of the commoner forms, viz, *Culex consobrinus,* *C. excitans, C. perturbans, C. posticatus, C. pungens, Prosophora ciliata,*

Anopheles punctipennis, and *A. quadrimaculata*, occur all over the country, from New England to Texas, and even to southern California. In almost any given locality in the United States, therefore, one would probably be able to find all of these eight species, with perhaps two or three additional ones.

The list which follows was drawn up by Mr. Coquillett, and embodies, in part, the results of his studies. It must be remembered that, after all, the material was scanty, since no one has taken the trouble to thoroughly collect mosquitoes. The list represents, however, a distinct and important advance on our former knowledge of these annoying creatures.

LIST OF THE MOSQUITOES OF THE UNITED STATES.

(A) *Species examined by D. W. Coquillett.*

Culex consobrinus Desv. 3 males, 18 females.

Synonyms: *Culex punctor* Kirby; *C. impatiens* Walk.; *C. pinguis* Walk.; *C. inornatus* Will. (the latter synonymy based on a study of one of Williston's cotype specimens).

Habitat: White Mountains, N. H.; Beverly, Mass., September 28 (Nat. Mus.); Catskill Mountains, Greene County, N. Y., 2,500 feet (Howard); Illinois, March 21, April 29, May 6, October 16 (Nason); St. Anthony Park, Minn., April, May, on snow (Lugger); Saskatchewan River, British America; South Dakota (Nat. Mus.); Lincoln, Nebr., May, September (Bruner); Colorado (Nat. Mus.); Los Angeles, Cal., February (Coquillett); Argus Mountains, Cal., April (Nat. Mus.); Santa Fé, N. Mex., July (Cockerell).

Culex excitans Walk. 3 males, 2 females.

Habitat: New Bedford, Mass. (Johnson); Lincoln, Nebr., May (Bruner); Santa Fé, N. Mex., July (Cockerell).

Culex excrucians Walk. 3 females.

Habitat: Ithaca, N. Y., July 14 (Comstock).

Culex fasciatus Fabr. 4 males, 2 females.

Synonyms: *Culex tæniatus* Wied.; *Culex mosquito* Desv. (non Arribalzaga).

Habitat: Georgia, August (Coquillett); Natchitoches, La., October 6 (Johnson); Isle of Pines, W. I. (Scudder); Kingston, Jamaica, July 13 (Johnson).

Culex impiger Walk. 14 males, 50 females.

Synonym: *Culex implacabilis* Walk.

Habitat: White Mountains, N. H.; Beverly, Mass., May 24, June 2 (Nat. Mus.); Ithaca, N. Y., July 9 and 17, August 28; Wilmuth, N. Y., June 10 (Comstock); Saskatchewan River, British America (Nat. Mus.); Minnesota (Lugger); London County, Va., Aug. 26 (Pratt); Tyrone, Ky., July 14 (Garman); Georgia (Nat. Mus.); Mesilla, N. Mex., (Cockerell); Isle of Pines, W. I. (Scudder); Portland, Jamaica (Johnson).

Culex perturbans Walk. 8 females.

Habitat: Lakeland, Md., August 8 (Pratt); Virginia, August 17 (Pergande); Tick Island, Fla., May 12 (Johnson); Texas (Nat. Mus.).

Culex posticatus Wied. 5 females.

Synonym: *Culex musicus* Say.

Habitat: Montgomery County, Pa., July 17 (Johnson); Texas (Nat. Mus.).

Culex pungens Wied. 25 males, 103 females.

Habitat: White Mountains, N. H.; Beverly, Mass., September 5; Cambridge, Mass., September 16 to November 5; Boston, Mass.; Baltimore, Md., November 5 (Nat. Mus.), November 26 (Lugger); Charlton Heights, Md., December 1

(Pratt); District of Columbia, January 30, March 5, May 6 and 15, June 28, July 11, August, October 10, 15, 25, and 31, November 4, 8, 13, 16, and 23, December 23 (Pergande); Ithaca, N. Y., May 29, July 17, August 28 (Comstock); Illinois (Nason); Minnesota (Lugger); Lincoln, Nebr., September 20 (Bruner); Lexington, Ky., November 10 (Garman); New Orleans, La., December 17 (Howard); San Antonio, Tex., May 5 (Marlatt); Georgia, August (Coquillett); Portland, Jamaica (Johnson).

Culex signifer Coq. 1 female.
Habitat: District of Columbia, June (Coquillett).

Culex stimulans Walk. 13 males, 54 females.
Habitat: White Mountains, N. H.; Beverly, Mass., June 2, July 9; Cambridge, Mass., May; Jamaica Plain, Mass., August 25 (Nat. Mus.); Baltimore, Md. (Lugger); Illinois, August 1, September 15, October 5 (Nason); Agricultural College, Mich. (Gillette); Saskatchewan River, British America (Nat. Mus.); Lincoln, Nebr. (Bruner); Colorado (Nat. Mus.); Ithaca, N. Y., June 13, 18, 29, July 14, August 28; Wilmuth, N. Y., June 10 (Comstock); Georgia (Nat. Mus.).

Culex tarsalis Coq. 1 male, 4 females.
Habitat: Argus Mountains, Cal., April; Folsom, Cal., July 3 (Nat. Mus.).

Culex triseriata Say. 3 females.
Habitat: White Mountains, N. H. (Nat. Mus.); Delaware County, Pa., June 12 (Johnson); Washington, D. C., May 5, Loudon County, Va. (Pratt).

Culex tæniorhynchus Wied. 1 male, 32 females.
(Not the *Culex tæniorhynchus* Wied. of Arribalzaga.)
Habitat: Maine, August; Beverly, Mass., June, September 15 (Nat. Mus.); Avalon, Anglesea, and Atlantic City, N. J., July 10 to 29 (Johnson); Far Rockaway, Long Island, N. Y., Aug. 30 (Howard); District of Columbia (Pergande); Georgia (Nat. Mus.); St. Augustine and Charlotte Harbor, Fla., July; Portland, Jamaica (Johnson).

Psorophora ciliata Fabr. 2 males, 29 females.
Habitat: Dorchester, Mass. (Nat. Mus.); Washington, D. C. (Chittenden); Westville, N. J., July 2 (Johnson); Illinois (Nason); Brooklyn Bridge, Ky., June 23 (Garman); Lincoln, Nebr., July, August (Bruner); Los Angeles, Cal. (Coquillett); San Diego, Tex., May 15 (Schwarz); Florida, July (Nat. Mus.).

Anopheles crucians Wied. 3 females.
Habitat: District of Columbia, April 27 (Pergande); Georgia (Nat. Mus.).

Anopheles punctipennis Say. 5 males, 13 females.
(Considered by Wiedemann to be the same species as his *Anopheles crucians*, but the two are certainly distinct.)
Synonym: *Culex hyemalis* Fitch (wrongly referred to *Anopheles quadrimaculata* in the Osten Sacken Catalogue).
Habitat: Castleton, Vt., February 1 (temperature 6° F.); Beverly, Mass., September 19, October 2; Cambridge, Mass., June 16, September 30, October 20 (Nat. Mus.); Charlton Heights, Md., March 31, November 17 (Pratt); District of Columbia, June 6, October 15, 25, and 31 (Pergande); Philadelphia, Pa., October 12 (Johnson); Ithaca, N. Y., April 17, August 28 (Comstock); Illinois, October 16 (Nason); Texas (Nat. Mus.); Mesilla, N. Mex. (Cockerell); Portland, Jamaica (Johnson).

Anopheles quadrimaculata Say. 3 males, 31 females.
Habitat: Berlin Falls, N. H., August (Nat. Mus.); Ithaca, N. Y., January, July 31, November 28 (Comstock); Lakeland, Md., August 8; Charlton Heights, Md., November 21 (Pratt); District of Columbia, July, October 15, November 2 and 14 (Pergande); Illinois, September 10, October 10 (Nason); St. Anthony Park, Minn., December 11 (Lugger); Tick Island, Fla., May 12 (Johnson); Texas (Nat. Mus.).

Megarhinus ferox Wied. 1 male.

   Habitat: District of Columbia, August 22 (Pergande).

Megarhinus rutilus Coq. 3 males, 5 females.

   Habitat: North Carolina; Georgiana, Fla. (Nat. Mus.).

Aëdes sapphirinus O. S. 1 female.

   Habitat: Ithaca, N. Y. (Comstock).

(B) *Species recorded from the United States, but not included in the material studied.*

Culex rubidus Desvoidy, Culicides. etc. Carolina.

Culex testaceus v. d. Wulp, Tijdschr. v. Entom., 2d ser., II, 128, Tab. III, f. 1. Wisconsin.

Culex incidens Thomson, Eugenie's Resa, etc., 443. California.

Culex territans Walker, Dipt. Saund., 428. United States.

Psorophora boscii Desvoidy, Culicides, etc. Carolina.

Anopheles annulimanus v. d. Wulp, Tijdschr. v. Entom., 2d ser., II, 129. Tab. III. f. 2. Wisconsin.

Anopheles ferruginosus Wiedemann, Auss. Zw., I, 12. New Orleans (Wied.); on the Mississippi (Say).

   Culex quinquefasciatus Say, Journ. Ac. Phil., III, 10, 2; Compl. Wr., II, 39. (Change of name by Wied.)

Anopheles maculipennis Meigen (European species, which also occurs in North America, according to Loew, Sillim. Journ., n. ser., Vol. XXXVII, 317).

Anopheles nigripes Staeger (European species, which also occurs in North America, according to Loew, Sillim. Journ., n. ser., Vol. XXXVII, 317).

Aëdes fuscus O. Sacken, Western Diptera, 191. Cambridge, Mass.

THE CAT AND DOG FLEA.

(*Pulex serraticeps* Gerv.)

Examination of many specimens of fleas sent to the Department in recent years shows that the species which commonly overruns houses during the damp summers, in our Eastern cities at least, is not, as many have supposed, the human flea (*Pulex irritans*), but the common cosmopolitan flea of the dog and the cat (*Pulex serraticeps*). There is widespread ignorance as to the transformations of this insect, and even the average entomologist is puzzled to know where to consult good figures of the different stages and a detailed account of the life history. The figures accompanying this article have been prepared to fill this want, and the following account of the transformations has been drawn up from notes made during the summer of 1895, at the request of the writer, by Mr. Pergande, of the division of entomology. The best two of the previously published articles are, one by Laboulbène, in the Annales de la Société Entomologique de France, 1872, pp. 267–273, Pl. XIII, and the other by W. J. Simmons, read before the Microscopical Society of Calcutta, March 5, 1888, and printed in The American Monthly Microscopical Journal for December, 1888, with no illustrations.[1]

---

[1] Ritzema has written an article on the natural history of the dog flea, which, however, could not be consulted by the writer.

Laboulbène describes carefully the pretty, oval, waxy white or opaque, porcelain-colored, smooth egg, which reaches 0.5 mm. in length. He describes the external appearance of the larvæ and recites their extremely rapid movements, which are made by means of the bristles with which they are furnished, and particularly by means of the tubercle and the hair-like spines below the head. He placed larvæ upon dust, with birds' feathers mixed with dried blood, upon which they developed perfectly. Others were put on the sweepings of a room, and developed just as well. Laboulbène at first believed that blood was necessary for the nourishment of the larvæ, the reddish-colored contents of the digestive tract making him think so; but he found they would flourish and complete their metamorphoses in sweepings in which there was no trace of blood. He concluded that all that has been said on *Pulex irritans* nourishing its young on dried blood is very problem-

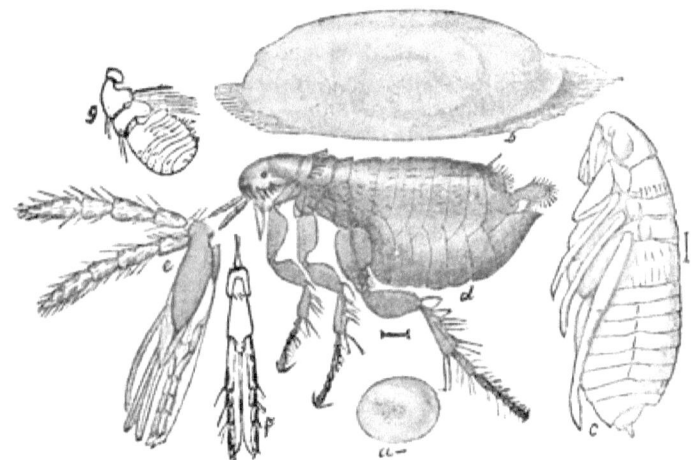

FIG. 5.— *Pulex serraticeps*; *a*, egg; *b*, larva in cocoon; *c*, pupa; *d*, adult; *e*, mouth-parts of same from side; *f*, labium of same from below; *g*, antenna of same—all enlarged (original).

atical. In his opinion the larvæ of the cat flea for the most part live upon the ground in spots where cats stay, and that they live in the dust in the cracks of the floor. The cocoon he described as ovoid, almost rounded, brown and granular, because it is covered with dust, delicate, but difficult to open, attached by one surface. It is about 2.5 mm. by 2.75 mm. The only statement in the article regarding the length of the different stages is to the effect that the pupal condition lasts from one to two weeks.

Mr. Simmons found the eggs upon a cloth upon which a dog had been sleeping, in the midst of a dust composed of fragments of cuticle, hairs, fibers, and pellets of dried blood, the last being probably the natural excreta of the fleas. In fifty hours most of the eggs hatched. The larvæ are described, and the statement is made that in seven days they began to spin their cocoons. They remained in the cocoons eight days, when the adults emerged, completing their transformations seventeen days after the eggs were deposited.

The eggs of the flea under consideration are deposited between the hairs of the infested animals, but are not fastened to them, so that when the animal moves about or lies down numbers of the eggs will be dislodged and drop to the ground or the floor or wherever the animal may be at the time. An easy way to collect them, therefore, is to lay a strip of cloth for the animal to sleep upon, and afterwards brush the cloth into a receptacle, in which the eggs will be found in numbers. Some difficulty was found in securing proper conditions of moisture to bring about successful rearing, and some detailed account of our experience will be of value to persons who desire to repeat the rearing in order to secure material for microscopic study, and will be at the same time suggestive as bearing on the conditions under which the insect will multiply in houses.

On June 27 a number of eggs were collected and placed in two glass vessels, one large and one small, each containing a layer of sand at the bottom, next a layer of sawdust, and on top of this a layer of rich soil. The eggs were placed between two layers of blotting paper on top of the soil. On June 29 fourteen of the eggs had hatched in the small vessel, and the larvæ had crawled at once down into the sawdust.

On July 1 some of the eggs were found to have hatched in the large vessel, and the alimentary canal of the larvæ was already brownish, indicating that they had been feeding to some extent and presumably upon the particles of dried blood collected with the eggs and placed with them between the layers of blotting paper. By July 11 all of these larvæ in both vessels had died, apparently without having cast a skin. They were very active during most of this period, crawling rapidly about when disturbed. Some were noticed to feed upon particles of peat which was placed with them. From some of these individuals fig. 6 was made. On the second antennal joint there was apparent a sensorial spot, and on or near the base of the antennæ were two small, slender, fleshy tubercles and a few granulations on each side, some distance behind the antennæ. At the base of the head above occurred a small, apparently well differentiated sclerite, as indicated in fig. 6, b, the purpose of which we can not surmise. Immediately behind it, on the anterior border of the first thoracic segment, is apparently a delicate sculpturing, indicating a thickening of the integument at this point. The posterior border of this segment is a somewhat similar, faintly indicated band. The first nine segments bear each four dorsal bristles and, on each side, one ventro-lateral bristle, near the posterior margin. The two following segments bear each six dorsal bristles and one ventro-lateral bristle, and the penultimate segment eight dorsal and one ventral bristle. These bristles become gradually longer toward the end of the body. The last segment is without long bristles, although there is a semicircular transverse row of numerous fine hairs and a small patch of still finer hairs on each of the anal lobes near the base of the anal prolegs, as shown in fig. 6, c.

On July 6 another lot of eggs was placed in each of the two different vessels. One lot was kept moist and the other dry, and both lots were provided with nothing but the particles of dried blood and a few crumbs of dry bread. On July 8 it was discovered that all of the eggs had hatched. Both vessels had been kept closed under a glass cover. Those between the layers of damp blotting paper had apparently not fed. Some were dead, having crawled up the sides of the vessel. Those in the dry receptable were very lively and had fed abundantly, so that the whole alimentary canal, from one end to the other, was dark brown.

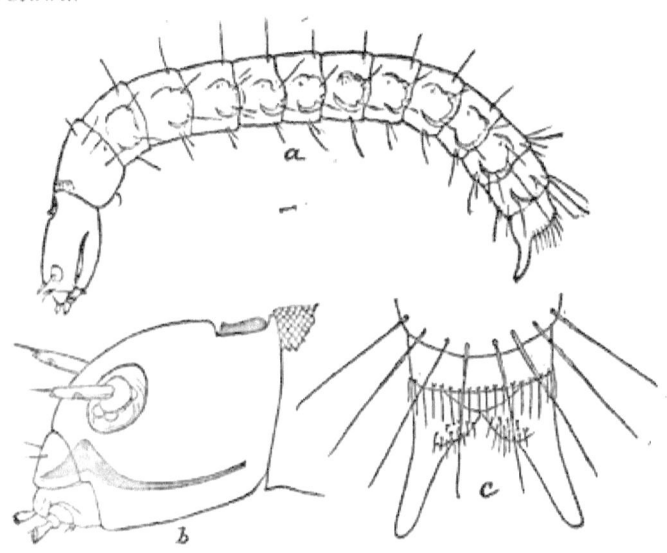

FIG. 6.—*Pulex serraticeps*; a, larva; b, head; c, anal end of same—greatly enlarged (original).

On July 9 the larvæ in the dry receptacle had cast the first skin, but upon careful examination were seen to agree perfectly with those of the first stage, except that they were larger. No trace of eyes could be found in either stage. The mandibles apparently bore four blunt teeth. At this date the larvæ kept in the moist receptacle had not cast a skin, and appeared almost colorless, having fed very little. In both vessels, however, all the larvæ were very active and ran about very briskly. Their movements when crawling recall those of many Tineid larvæ. Ten individuals of the second stage were removed to another vessel to see whether they would cast a second skin.

On July 10 all of the larvæ in the original moist vessel died. Those in the dry vessel, which had been fed with bread crumbs, were still growing nicely, and were very active. By July 15 all the larvæ which had been transferred, to watch for further molts, had died without molting. They either stuck to the crumbs, which were rather greasy, or to the sides of the glass, which had also become somewhat greasy. On the same date the larvæ in the dry vessel, from which these ten had been removed, commenced to spin up. Many were restlessly running about in search of suitable places for spinning, and some had even reached the top of the blotting paper. A thin layer of gray cotton was placed between the two blotting papers to give them suitable spinning places. The eggs hatched in two days, having been kept dry all the time. The first skin was cast two days after hatching, and the beginning of spinning occurred eight days after hatching.

By July 19 no more specimens had spun up and many had died. The receptacle seemed to be too dry and too hot, and the blotting paper was somewhat moistened. But one pupa was found, which was that of a larva which began to spin July 15. July 21 no others had spun up, although they were still very lively. The pupa had become brownish. July 22 the adult flea issued over night and escaped through the covering. From that time until July 29 no more spun cocoons, and many of them died. On July 30 one of the survivors commenced to spin, twenty-four days after hatching. The cocoon is delicate, white in color, and is very well shown at fig. 5.

On August 2 this larva, which commenced to spin July 21, changed to pupa. On August 6 it was still white in color, becoming somewhat yellowish on the 7th and quite brown on the 8th. On the 9th the adult flea was found to have issued overnight. The pupa state, therefore, lasted about eight days, and it is to be noted that the pupa remains white until shortly before the emergence of the adult. It was supposed that the pupa stage in this instance was longer than usual, on account of the fact that the larval stage was so very much longer than in the first instance.

On July 11 another series of experiments was started, in order to gauge the variation in the duration of the stages and settle the question of the number of larval molts. Eggs collected on this date hatched July 13. On July 16, of fifteen larvæ eleven had cast the first skin. On July 18 five specimens cast the second skin. July 19 the weather was extremely warm and a number of the larvæ died. July 20 the heat continued, and more died. On July 23 seven larvæ which had cast the first skin remained; one of them had begun to spin up. There were on the morning of this date three cast skins in the receptacle, so that there are apparently three molts. In this final state the bristles have become longer and the mandibles have two teeth at the apex. The remaining four were carried on until August 8, when the last one died, none of them having succeeded in casting a third skin. Of the entire lot, but one was reared to the pupa state, and this pupa was preserved in alcohol for drawing. The record of this advanced specimen shows three molts, and that it began to spin eight days after hatching. The average of the others shows that the eggs hatch in from two to four days and that some of the larvæ cast their first skin three to four days later, and a second skin two to six days later.

On July 15 another series was begun. The eggs collected on this date began to hatch on the 17th and all had hatched by the morning of the 18th. July 21 some of them had cast the first skin.

August 1 the first one spun up; August 3, two more; August 6, two more. At this date the first one which constructed its cocoon turned brown. August 7 one full-grown larva transformed to pupa without spinning a cocoon. August 12 the first adult emerged. A summary for this lot shows that the eggs hatch in from two to four days and that the larvæ cast the first skin from five to seven days later. Some spun up sixteen to twenty days after hatching, and the imago appeared six days later.

Observation of these last two lots shows that the larvæ are very apt to die if kept too dry or too moist. They also need plenty of air.

July 20 another series was begun. Eggs collected on this date hatched the following day. July 24 the first skin was cast; July 26, in one case a second skin was cast. July 27 three more cast a second skin, and on this date one individual spun its cocoon. July 29 three more began to spin; on July 30 many more. On July 30 the first one that began to spin was found to have changed to pupa. August 2 many cocoons were found. Some of the larvæ, disturbed while spinning, left the incomplete cocoon and transformed to pupa outside of it. Most of the advanced specimens were placed in alcohol, and it was not until August 14 that an adult was allowed to emerge.

This series of observations showed that the eggs hatched about one day after being placed in the vessels. The larvæ cast their first skin in from three to seven days, and their second skin in from three to four days. They commenced spinning in from seven to fourteen days after hatching, and the imago appeared five days later.

From these observations it appears that in summer at Washington many specimens will undergo their transformations quite as rapidly as Mr. Simmons found to be the case at Calcutta, and that an entire generation may develop in little more than a fortnight; also that an excess of moisture is prejudicial to the successful development of the insect and that in the same way the breeding place must not be too dry. The little particles of blood found among the eggs on the cloth upon which the infested animal has slept are probably the excrement of the adult fleas. This substance in itself, together with what vegetable dust is found in the places where these larvæ rear themselves, suffices for the larval food.

REMEDIES.

Flea larvæ will not develop successfully in situations where they are likely to be disturbed. That they will develop in the dust in the cracks in floors which are not frequently swept has been observed by the writer. The overrunning of houses in summer during the temporary absence of the occupants is undoubtedly due to the development of a brood of fleas in the dust in the cracks of the floor from eggs which have been dropped by some pet dog or cat. This overrunning is more liable to occur in moist than in excessively dry summer weather, and it is more likely to occur during the absence of the occupants of the house, for the reason that the floors do not, under such circumstances, receive their customary sweeping. The use of carpets or straw mattings, in our opinion, favors their development under the circumstances above mentioned. The young larvæ are so slender and so active that they readily penetrate the interstices of both sorts of coverings and find an abiding place in some crack where they are not likely to be disturbed.

That it is not difficult to destroy this flea in its early stages is shown by the difficulty we have had in rearing it; but to destroy the adult fleas is another matter. Their extreme activity and great hardiness render any but the most strenuous measures unsuccessful. In such cases we have tried a number of the ordinarily recommended remedies in vain. Even the persistent use of California buhach and other pyrethrum powders, and, what seems still stranger, a free sprinkling of floor matting with benzine, were ineffectual in one particular case of extreme infestation. In fact, it was not until all the floor mattings had been taken up and the floor washed down with hot soapsuds that the flea pest abated. In another case, however, the writer found that a single application of California buhach, freely applied, was perfectly successful; and in a third case a single thorough application of benzine also resulted in perfect success. The pyrethrum application was made in a Brooklyn (N. Y.) house, and the benzine application in a Washington residence. The frequently recommended newspaper remedy of placing a piece of raw meat in the center of a piece of sticky fly paper has been thoroughly tried by the writer, without the slightest success. As a palliative measure, however, the plan adopted by Professor Gage in the McGraw Building of Cornell University, and described at length on page 422 of Vol. VII, Insect Life, may be worth trying. It will be remembered that Professor Gage tied sheets of sticky fly paper, with the sticky side out, around the legs of the janitor of the building, who then for several hours walked up and down the floor of the infested room, with the result that all or nearly all of the fleas jumped on his ankles, as they will always do, and were caught by the fly paper.

In his recent summary of the described fleas (Canadian Entomologist, August, 1895, pp. 221-222) Mr. C. F. Baker shows that there are forty-seven valid species, which attack all sorts of warm-blooded animals. The species which we have just considered (*Pulex serraticeps* Gervais) is, as he states, the common cat and dog flea, well known over all parts of the world. Mr. Baker further states that, "besides the various wild cats and dogs, it has been reported from *Herpestes ichneumon* (Pharaoh's rat), *Fœtorius putorius* (common polecat of

Europe). *Hyæna striata* (striped hyena), *Lepus timidus* (common hare), and *Procyon lotor* (raccoon). It is also said to occasionally sip human blood [*sic!*]. I have specimens from various parts of North America, and also from Europe." Many unfortunate inhabitants of New York, Philadelphia, Washington, and Baltimore during the past few summers will be able to verify Mr. Baker's statement that the species occasionally sips human blood! This species may be distinguished at a glance from the so-called human flea (*Pulex irritans*) by the fact that the latter species does not possess the strong recurved spines on the margin of the head, which show so distinctly in fig. 5.

CHAPTER II.

THE BEDBUG AND CONE-NOSE.

By C. L. Marlatt.

THE BEDBUG.

(*Cimex lectularius* Linn.)

This disgusting human parasite, the very discussion of which is tabooed in polite society, is practically limited to houses of the meaner sort, or where the owners are indifferent or careless, or to hostelries not always of the cheaper kind. The careful housekeeper would feel it a signal disgrace to have her chambers invaded by this insect, and, in point of fact, where ordinary care and vigilance are maintained the danger in this direction is very slight. The presence of this insect, however, is not necessarily an indication of neglect or carelessness, for,

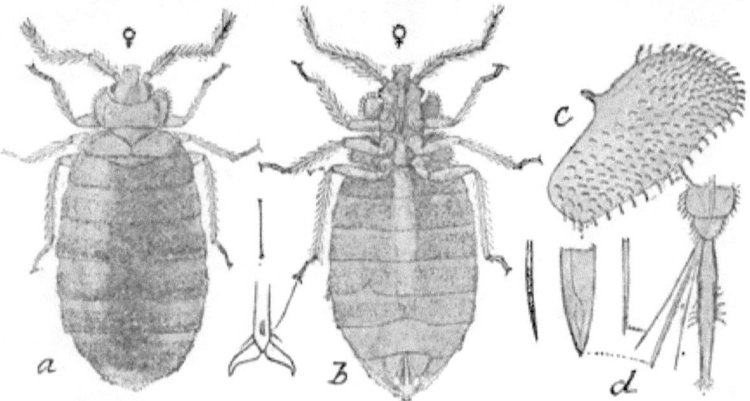

Fig. 7.—*Cimex lectularius:* a, adult female, gorged with blood; b, same, from below; c, rudimentary wing-pad; d, mouth-parts—all enlarged (original).

little as the idea may be relished, it may often gain access in spite of the best of care and the adoption of all reasonable precautions. It is very apt to get into the trunks and satchels of travelers, and may thus be introduced into homes. Unfortunately, also, it is quite capable of migrating from one house to another, and will often continue to come from an adjoining house, sometimes for a period of several months, gaining entrance daily. Such migration is especially apt to take place if the human inhabitants of an infested house leave it. With the failure of their usual source of food, the migratory instinct is developed, and escaping through windows, they pass along walls, water pipes, or gutters, and thus gain entrance into adjoining houses. It is expedient, therefore, to consider this insect, unsavory as the subject may be, since, as shown, it may be anyone's misfortune to have his premises temporarily invaded.

As with nearly all the insects associated with man, the bedbug has had the habits now characteristic of it as far back as the records run. It was undoubtedly of common occurrence in the dwellings of the ancient peoples of Asia. The Romans were well acquainted with it, giving it the name Cimex. It was supposed by Pliny (and this was doubtless the common belief among the Romans) to have medicinal properties, and it was recommended, among other things, as a specific for the bites of serpents. It is said to have been first introduced into England in 1503, but the references to it are of such a nature as to make it very probable that it had been there long previously. Two hundred and fifty

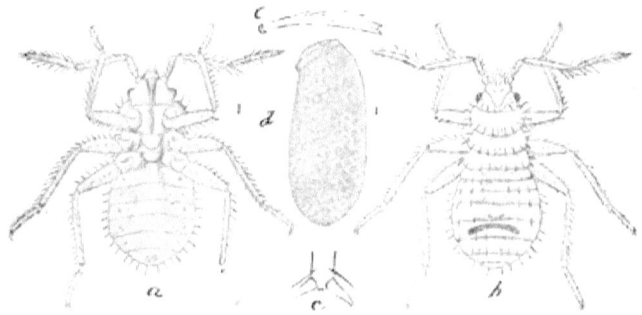

Fig. 8.—*Cimex betularius*. Egg and newly hatched larva of bedbug: *a*, larva from below; *b*, larva from above; *c*, claw; *d*, egg; *e*, hair or spine of larva—greatly enlarged; natural size of larva and egg indicated by hair lines (original).

years later it was reported to be very abundant in the seaport towns, but was scarcely known inland. It has been inferred that the following reference from the old English Bible of 1551 is to this insect: "Thou shalt not nede to be afraid for eny Bugges by night" (Psalm XCI, 5).

One of the old English names was "wall-louse." It was afterwards very well known as the "chinch," which continued to be the common appellation for it until within a century or two, and is still used in parts of this country. The origin of the name "bedbug" is not known, but it is such a descriptive one that it would seem to have been very naturally suggested. Almost everywhere there are local names for this parasite, as, for illustration, around Boston they are called "chintzes" and "chinches," and from Baltimore comes the name "mahogany flat," while in New York they are styled "red coats."

The bedbug has accompanied man wherever he has gone. Vessels are almost sure to be infested with it. It is not especially limited by cold, and is known to occur well north. It probably came to this country with the earliest colonists, at least Kalm, writing in 1748-49, stated that it was plentiful in the English colonies and in Canada, though unknown among the Indians.

The bedbug belongs to the order Hemiptera, which includes the true bugs or piercing insects, characterized by possessing a piercing and sucking beak. The bedbug is to man what the chinch bug is to grains or the squash bug to cucurbs. Like nearly all the insects parasitic on animals, however, it is degraded structurally, its parasitic nature and the slight necessity for extensive locomotion having resulted, after many ages doubtless, in the loss of wings and the assumption of a comparatively simple structure. The wings are represented by the merest rudiments, barely recognizable pads, and it lacks the simple eyes or ocelli of most other true bugs. In form it is much flattened, obovate, and in color is rust red, with the abdomen more or less tinged with black. The absence of wings is a most fortunate circumstance, since otherwise there would be no safety from it even for the most careful and thorough of housekeepers. Some slight variation in length of wing pads has been observed, but none with wings showing any considerable development have ever been found.

A closely allied species is a parasitic messmate in the nests of the common barn or eaves swallow in this country, and it often happens that the nests of these birds are fairly alive with these vermin. The latter not infrequently gain access to houses, and cause the housekeeper considerable momentary alarm. At least three species occur also in England, all very closely resembling the bedbug. One of these is found in pigeon cotes, another in the nests of the English martin, and a third in places frequented by bats. What seems to be the true bedbug, or at least a mere variety, also occurs occasionally in poultry houses.[1]

The most characteristic feature of the insect is the very distinct and disagreeable odor which it exhales, an odor well known to all who have been familiar with it as the "buggy" odor. This odor is by no means limited to the bedbug, but is characteristic of most plant bugs also. The common chinch bug affecting small grains and the squash bugs all possess this odor, and it is quite as pungent with these plant-feeding forms as with the human parasite. The possession of this odor, disagreeable as it is, is, after all, a most fortunate circumstance, as it is of considerable assistance in detecting the presence of these vermin. The odor comes from glands, situated in various parts of the body, which secrete a clear, oily, volatile liquid. The possession of this odor is certainly, with the plant-feeding forms, a means of protection against insectivorous birds, rendering these insects obnoxious or distasteful to their feathered enemies. With the bedbug it is probably an illustration of a very common phenomenon among animals, the persistence of a characteristic which is no longer of any especial value to the possessor of it. The natural enemies of true bugs, against which this odor serves

[1] Insect Life, Vol. VI, p. 166, Osborn.

as a means of protection, in the conditions under which the bedbug lives, are kept away from it, and the roach, which will be shown later to feed on bedbugs, is evidently not deterred by the odor, while the common house ant, which will also attack the bedbug, seems not to find this odor disagreeable.

The bedbug is thoroughly nocturnal in habit and displays a certain degree of wariness and caution, or intelligence, in its efforts at concealment during the day. It thrives particularly in filthy apartments and in old houses which are full of cracks and crevices in which it can conceal itself beyond easy reach. It usually leaves the bed at the approach of daylight to go into concealment either in cracks in the bedstead, if it be one of the old wooden variety, or behind wainscoting, or under loose wall paper, where it manifests its gregarious habit by collecting in masses together. The old-fashioned heavy wooden bedsteads are especially favorable for the concealment and multiplication of this

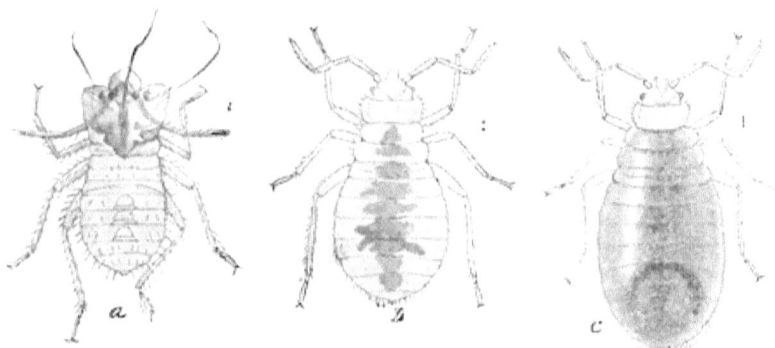

Fig. 9.—*Cimex lectularius:* a, first larval skin shed at first moult; b, second larval stage taken immediately after emerging from a; c, same after first meal, distended with blood (original).

insect, and the general use in later years of iron and brass bedsteads has very greatly facilitated its eradication. They are not apt to be very active in winter, especially in cold rooms, and ordinarily hibernate in their places of concealment.

The bedbug, though normally feeding on human blood, seems to be able to subsist for a time at least on much simpler food, and in fact the evidence is pretty conclusive that it is able to get more or less sustenance from the juices of moistened wood, or the moisture in the accumulations of dust, etc., in crevices in flooring. No other explanation would seem to account for the fact that houses long unoccupied are found, on being reinhabitated, to be thoroughly stocked with bedbugs.

There is a very prevalent belief among the old settlers in the West that this insect normally lives on dead or diseased cottonwood logs, and is almost certain to be abundant in log houses of this wood. This belief was recently voiced by Capt. S. M. Swigert, U. S. A., who reports that it often occurs in numbers under the bark of dead trees of cottonwood (*Populus monilifera*), especially along the Big and Little Horn rivers in Montana.

The origin of this misconception—for such it is—so far as the out-of-door occurrence is concerned, is probably, as pointed out by Professor Riley, from a confusion of the bedbug with the immature stages of an entirely distinct insect (*Aradus* sp.) which somewhat resembles the former and often occurs under cottonwood bark. In houses, green or moist cottonwood logs or lumber may actually furnish sustenance in the absence of human food. The bedbug is, however, known to be able to survive for long periods without food, specimens having been kept for a year in a sealed vial, with absolutely no means of sustenance whatever, and in unoccupied houses it can undoubtedly undergo fasts of extreme length. Individuals obtained from eggs have been kept in small sealed vials in this office for several months, remaining active and sprightly in spite of the fact that they had never taken any nourishment whatever.

Extraordinary stories are current of the remarkable intelligence of this insect in circumventing various efforts to prevent its gaining access to beds. Most of these are undoubtedly exaggerations, but the inherited experience of many centuries of companionship with man, during which the bedbug has always found its host an active enemy, has resulted in a knowledge of the habits of the human animal and a facility of concealment, particularly as evidenced by its abandoning beds and going often to distant quarters for protection and hiding during daylight, which indicate considerable apparent intelligence.

The bite of the bedbug is decidedly poisonous to some individuals, resulting in a slight swelling and disagreeable inflammation. To such persons the presence of bedbugs is sufficient to cause the greatest uneasiness, if not to put sleep and rest entirely out of the question. With others, however, who are less sensitive, the presence of the bugs may not be recognized at all, and, except for the occasional staining of the linen by a crushed individual, their presence might be entirely overlooked. The inflammation experienced by sensitive persons seems to result merely from the puncture of the skin by the sharp piercing setæ which constitute the puncturing element of the mouth parts, as there seems to be no secretion of poison other than the natural fluids of the mouth.

The biting organ of the bedbug is exactly like that of other hemipterous insects. It consists of a rather heavy, fleshy under lip (the only part ordinarily seen in examining the insect), within which lie four thread-like hard filaments or setæ which glide over each other with an alternating motion and pierce the flesh. The blood is drawn up through the beak, which is closely applied to the point of puncture, and the alternating motion of these setæ in the flesh causes the blood to flow more freely. The details of the structure of the beak are shown in the accompanying sketch (fig. 7, *d*). In common with other insects which attack men, it is entirely possible for these pests to be transmitters of contagious diseases.

Like its allies, the bedbug undergoes an incomplete metamorphosis, the young being very similar to their parents in appearance, structure, and in habit. The eggs are white oval objects, having a little projecting rim around one edge, and are laid in batches of from one-half dozen to fifty in cracks and crevices where the bugs go for concealment. The eggs hatch in a week or ten days, and the young escape by pushing the lid within the projecting rim from the shell. At first they are yellowish white, nearly transparent, the brown color of the more mature insect increasing with the later molts. During the course of development the skin is shed five times, and with the last molt the minute wing pads characteristic of the adult insect make their appearance. A period of about eleven weeks has been supposed to be necessary for the complete maturity of this insect, but we have found this period subject to great variation, depending on warmth and food supply. Breeding experiments conducted at this office indicate, under most favorable conditions, a period averaging eight days between moltings and between the laying of the eggs and their hatching, giving about seven weeks as the period from egg to adult insect. Some individuals under the same conditions will, however, remain two to three weeks between moltings, and without food as already shown they may remain unchanged for an indefinite time. Ordinarily but one meal is taken between molts, so that each bedbug must puncture its host five times before becoming mature and at least once afterwards before it again develops eggs. They are said to lay several batches of eggs during the season, and are extremely prolific, as occasionally realized by the housekeeper, to her chagrin and embarrassment.

REMEDIES.

The bedbug, on account of its habits of concealment, is usually beyond the reach of powders, and the ordinary insect powders, such as pyrethrum, are of practically no avail against it. If iron or brass bedsteads are used the eradication of the insect is comparatively easy. With large wooden bedsteads, furnishing many cracks and crevices into which the bugs can force their flat, thin bodies, their extermination becomes a matter of considerable difficulty. The most practical way to effect this end is by very liberal applications of benzine or kerosene or any other of the petroleum oils. These must be introduced into all crevices with small brushes or feathers, or by injecting with small syringes. Corrosive sublimate is also of value, and oil of turpentine may be used in the same way. The liberal use of hot water wherever it may be employed without danger to furniture, etc., is also an effectual method of destroying both eggs and active bugs. Various bedbug remedies and mixtures are for sale, most of them containing one or the other of the ingredients mentioned, and they are frequently of value. The great desideratum, however, in a case of this kind, is a daily inspection of beds and bedding and of all crevices and locations about the premises where these vermin may have gone for concealment. A vigorous campaign should, in the course of a week or so at the outside, result in the extermination of this very obnoxious and embarrassing pest. In the case of rooms containing books or where liquid applications are inadvisable, a thorough fumigation with brimstone is, on the authority of Dr. J. A. Lintner, New York State entomologist, an effective means of destruction. He says:

Place in the center of the room a dish containing about 4 ounces of brimstone, within a larger vessel, so that the possible overflowing of the burning mass may not injure the carpet or set fire to the floor. After removing from the room all such metallic surfaces as might be affected by the fumes, close every aperture, even the keyholes, and set fire to the brimstone. When four or five hours have elapsed, the room may be entered and the windows opened for a thorough airing.

The fact that the bedbug has a very effective enemy in the common house cockroach has already been alluded to, and is particularly described in the chapter on the cockroach. Another common insect visitor in houses, and a very annoying one also to the careful housekeeper, the little red ant (*Monomorium pharaonis*), is also known to be a very active and effective enemy of the bedbug. Mr. Theo. Pergande, of this office, informs me that during the late war, when he was with the Union army, he occupied at one time barracks at Meridian, Miss., which had been abandoned by the Southern troops some time before. The premises proved to be swarming with bedbugs; but very shortly afterwards the little red house ant discovered the presence of the bedbugs and came in in enormous numbers, and Mr. Pergande witnessed the very interesting and pleasing sight of the bedbugs being dismembered or carried away bodily by these very minute ants, many times smaller than the bugs which they were handling so successfully. The result was that in a single day the bedbug nuisance was completely abated. The liking of red ants for bedbugs is confirmed also by a correspondent writing from Florida (F. C. M. Boggess), who goes so far as to heartily recommend the artificial introduction of the ants to abate this bug nuisance. (Insect Life, Vol. VI, p. 340.) Bedbugs and other household insects, however, are not of the sort which it is convenient or profitable to turn over to their natural enemies in the hope that eradication by this means will follow, and the fact of their being preyed upon by other insects furnishes no excuse to the housekeeper for not instituting prompt remedial measures.

THE BLOOD-SUCKING CONE-NOSE.

(*Conorhinus sanguisuga* Lec.)

Somewhat allied to the bedbug in habit is another true bug, *Conorhinus sanguisuga*, bearing the very descriptive and appropriate popular name of the "blood-sucking cone-nose," or sometimes called the

Texas or Mexican bedbug, or simply the big bedbug. Until recently it has been a rare visitant in houses, and is still practically unknown in Eastern cities, but in country places, particularly in the Mississippi Valley, is now often found in bedrooms, and its bite is very severe and painful, resulting in much more pronounced swelling and inflammation than in the case of the bedbug.

The cone-nose belongs to the group of true bugs which includes predaceous species, or those which normally feed on other insects rather than on plant juices. The members of the genus Conorhinus are mostly South American, and, on the authority of Burmeister, have the habit in the adult state of living, in part at least, on the blood of mammals. The normal food of our species is, however, unquestionably other insects, and its liking for human blood is evidently a habit of recent acquisition and limited to the full-grown insect, and probably only a small percentage of these ever taste blood. Miss Bertha Kimball (Trans. Kans. Acad. Sci., Vol. XXIV, p. 128, 1896) reports that they are often found in poultry houses, and that when abundant they attack horses in barns, and probably other domestic animals.

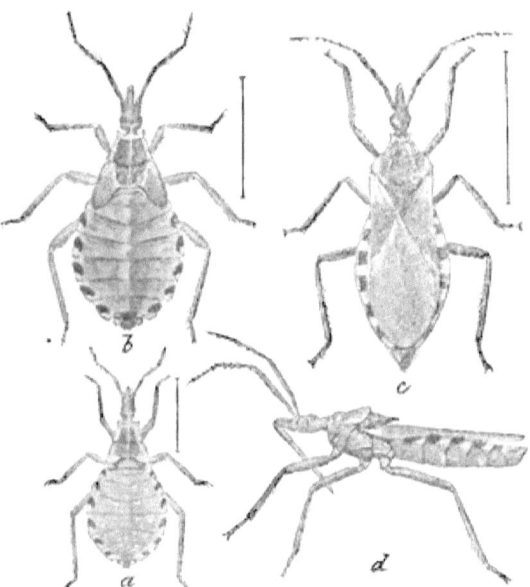

Fig. 10.—*Conorhinus sanguisuga:* a. first pupal stage; b, second pupal stage; c, adult bug; d, same, lateral view—all enlarged to same scale (original).

In houses it has been found with bedbugs, and will unquestionably feed upon them, especially if it can secure specimens already charged with human blood, and it has been actually observed eating what was taken to be a young roach. In captivity Miss Kimball has succeeded in feeding both young and adults on house flies. That the blood-taking habit may be easily acquired is shown by the fact that many common plant bugs, if captured, will pierce the flesh, and several of the species which are attracted to light at night and settle on one's hand will pierce the skin and fill themselves with blood.

The accompanying figures of this insect represent the egg, newly hatched larva, and last larval stage, drawn to the same scale (fig. 11), and the pupal stages and the adult, also drawn to a scale, but less magnified than the others (fig. 10). The eggs and young larvæ have recently been described for the first time by Miss Kimball (l. c.), and this summer a large number of specimens in all stages were received from the West, from which the accompanying figures were made. From these specimens many eggs were obtained, and later, larvæ.

The cone-nose is a rather large insect, measuring an inch in length and characterized by a flattened body and very narrow, pointed head and short, strong beak. In color it is dark brown, with the light areas indicated in the figure pinkish. Its "buggy" odor is even more intense than that of the bedbug. It is a night flyer and is attracted into open

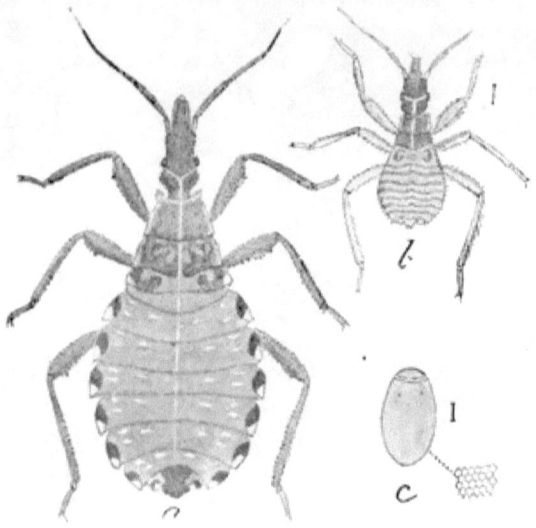

FIG. 11 —*Conorhinus sanguisuga*: *a,* larva, second stage; *b,* newly hatched larva; *c,* egg with sculpturing of surface shown at side—all enlarged to same scale—(original).

windows by lights. It conceals itself during the day under any loose object, often leaving beds which it may have frequented during the night. The adult is not apt to take flight, but can run rather swiftly.

The eggs are white, changing to yellow and pink before hatching, and of the peculiar shape indicated in the illustration. The young hatch within twenty days. There are at least two larval stages (fig. 11. *a, b*) and two pupal stages (fig. 10. *a, b*), the latter characterized by the presence of distinct wing pads. In all these stages the insect is active and predaceous. The eggs are normally deposited and the early stages are undoubtedly passed out of doors, the food of the immature forms being other insects. The eggs which may be dropped indoors must fail normally to mature adults, and in fact immature specimens are rarely found indoors, and the wingless and rather sluggish larvæ and pupæ would have little opportunity of reaching the higher animals under any circumstances. It winters, both in the partly grown and the adult state, often under bark of trees or in any similar protection, and only in its nocturnal spring and early summer flights does it become an enemy of man in the effort to gratify its taste for human blood.

This insect is particularly abundant and usually enters houses in early spring (April and May), sometimes in considerable numbers, and seems to be decidedly on the increase in the region which it particularly affects—the plains region from Texas northward and westward. A correspondent in Indian Territory reported having in the course of a short while killed upward of a dozen. They were usually found in the bed or near by, and their connection with the injury was often very plainly evident by their being found turgid with blood.

The common California species closely resembles in appearance and habits the one named at the head of this section, but is a distinct species and apparently undescribed. The local name in California for this insect is "monitor bug."

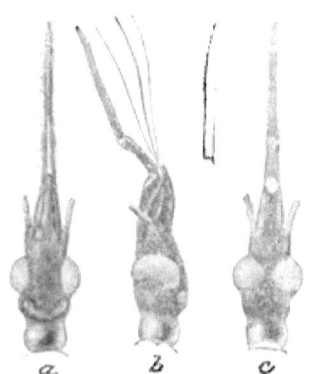

FIG. 12.—*Conorhinus sanguisuga*: a, head, showing beak; b, same, from the side, with piercing setæ removed from sheath and with tip of one of them enlarged; c, same, from below—much enlarged (original).

The results of the bite of the cone-nose on the human subject vary a good deal with the susceptibility of the person bitten, but are often of a very serious and alarming character. The piercing of the skin is evidently accompanied by the injection of some poisonous liquid or venom, making a sore, itching wound, accompanied with a burning pain lasting sometimes from two to four days, and often associated with swellings, which may extend over a good deal of the body. That there is a specific poison injected is indicated rather conclusively by the very constant and uniform character of the symptoms in nearly all cases of bites by this insect. It has, however, been suggested that the very serious results which sometimes follow its bite may be due to the fact that it has previously thrust its beak into some decaying animal matter, causing a certain amount of blood poisoning in the patient. This theory has support in the facts stated by the late J. B. Lembert, of California, who says that he has noticed that the species of Conorhinus occurring on the Pacific Slope is attracted by carrion. Mr. Lembert described the effect on himself of a sting by this insect on the middle toe of the left foot. Following the sting an itching sensation extended up the leg, large blotches manifesting themselves on the upper part of the limb and extending up to the hands and arms. His lips swelled, and the itching and swelling extended over the head, and he was also much nauseated. The itching abated after four or five hours, but the swelling did not go down until the next day. A correspondent, writing to Prof. J. W.

Toumey, describes similar results from a sting from one of these insects in Arizona. The patient, a woman, broke out over the body and limbs with red blotches or welts, like a severe case of measles, from a sting on the shoulder. Bathing with sweet oil soon reduced the dangerous symptoms, which were accompanied with severe headache and nausea. Similar results following the puncture of this insect have been reported from Indian Territory, Kansas, and elsewhere. Miss Kimball (l. c.) says that some relief from the effects of the bites of this insect is afforded by camphor, ammonia, and the ordinary remedies for insect stings.

To attempt to control the out-of-door multiplication of this insect is manifestly out of the question, and in the screening of the entrances of houses or chambers is the only practical method of protection. It hardly needs stating that all examples found should be promptly killed.

CHAPTER III.

HOUSE FLIES, CENTIPEDES, AND OTHER INSECTS THAT ARE ANNOYING RATHER THAN DIRECTLY INJURIOUS.

By L. O. Howard and C. L. Marlatt.

HOUSE FLIES.

(*Musca domestica, et al.*)

In common parlance there is but one house fly, although a number of species are in the habit of entering houses and cause more or less annoyance. The most abundant form is the house fly proper (*Musca domestica* Linn.). It is a medium-sized, grayish fly, with its mouth parts spread out at the tip for sucking up liquid substances. It breeds in manure and dooryard filth and is found in nearly all parts of the

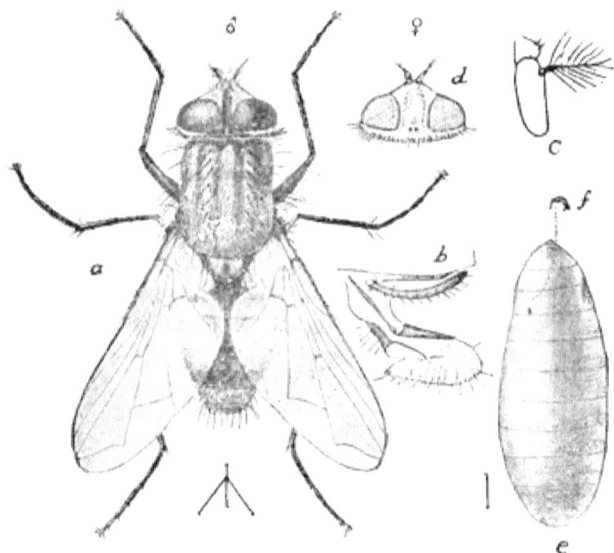

FIG. 15.—*Musca domestica*: *a*, adult male; *b*, proboscis and palpus of same; *c*, terminal joints of antenna; *d*, head of female; *e*, puparium; *f*, anterior spiracle—all enlarged (original).

world. On account of the conformation of its mouth parts, the house fly can not bite, yet no impression is stronger in the minds of most people than that this insect does occasionally bite. This impression is due to the frequent occurrence in houses of another fly (*Stomoxys calcitrans*), which may be called the stable fly, and which, while closely resembling the house fly (so closely, in fact, as to deceive anyone but an

49

entomologist), differs from it in the important particular that its mouth parts are formed for piercing the skin. It is perhaps second in point of abundance to the house fly in most portions of the Northeastern States.

A third species, commonly called the cluster fly (*Pollenia rudis*), is a very frequent visitant of houses, particularly in the spring and fall. This fly is somewhat larger than the house fly, with a dark-colored, smooth abdomen and a sprinkling of yellowish hair. It is not so active as the house fly, and particularly in the fall is very sluggish. At such times it may be picked up readily, and is very subject to the attacks of a fungus disease which causes it to die upon window panes surrounded by a whitish efflorescence. Occasionally this fly occurs in houses in such numbers as to cause great annoyance, but such occurrences are comparatively rare.

A fourth species is another stable fly known as *Cyrtoneura stabulans*, and a fifth, rather commoner than the last, is the so-called bluebottle fly (*Calliphora erythrocephala*). This insect is also called the blowfly or meat fly, and breeds in decaying animal material. Another species, about the size of the bluebottle, which breeds abundantly in cow-dung and is also found in houses, although usually in less numbers than the others, is also commonly called the bluebottle or green-bottle fly (*Lucilia cæsar*).

There is still another species, smaller than any of those so far mentioned, which is known to entomologists as *Homalomyia canicularis*, sometimes called the small house fly. It is distinguished from the ordinary house fly by its paler and more pointed body and conical shape. The male, which is much commoner than the female, has large pale patches at the base of the abdomen, which are translucent. When seen on a window pane the light shines through that part of the body. Not much complaint would be made of house flies were the true house fly a nonexistent form. Under ordinary circumstances it far outnumbers all other species in houses. Common and widespread as this species is, there is very general ignorance, as with many other extremely common insects, as to its life history and habits outside of the adult stage. Writing in 1873, Dr. A. S. Packard[1] showed that no one in this country had up to that time investigated its habits, and that even in Europe but little attention had been given to it. He showed that the habits were mentioned in only three works, one of which was published during the present century, with figures so poor and inadequate as to be actually misleading. De Geer (1752) showed that the larva lives in warm and humid dung, but did not say how long it remains in the different stages. Bouché (1834) states that the larva lives in horse and fowl's dung, especially when warm; he did not, however, give the length of the larval state.

---

[1] On the Transformations of the Common House Fly, with Notes on Allied Forms. Proc. Boston Soc. Nat. Hist., Vol. XVI, 1874, p. 136.

Dr. Packard studied the species with some care, and obtained large numbers of the eggs by exposing horse manure. He carefully followed the transformations of the insect, and gave descriptions of all stages. He found the duration of the egg state to be twenty-four hours, the duration of the larval state five to seven days, and of the pupal state five to seven days. The period from the time of hatching to the exclusion of the adult, therefore, occupies, according to Packard, from ten to fourteen days. His observations were made at Salem, Mass.

As is quite to be expected, as we go further south the house fly becomes more numerous and more troublesome. The number of generations annually increases as the season becomes longer, and with the warm climate the development of the larva becomes more rapid. A few rearing experiments were made in this office during the summer of 1895, and it was unexpectedly found that the house fly is a difficult insect to rear in confinement. Buzzing about everywhere, and apparently living with ease under the most adverse conditions, it is nevertheless, when confined in the warm season of the year to a small receptacle, not at all tenacious of life. It results from this fact, for example, that it is almost impossible to ascertain the length of the life of the house fly in the adult condition. On June 26 a small quantity of fresh horse manure was exposed in a fly-infested room for a few minutes. The flies

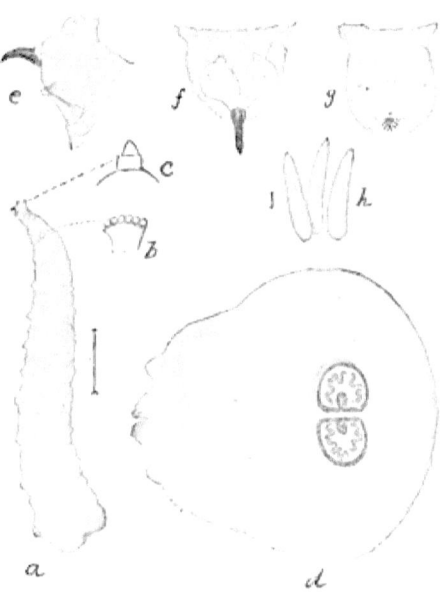

Fig. 14. *Musca domestica*: a, full-grown larva; b, one of its anterior spiracles; c, side view of head; d, hind end of body showing anal spiracles; e, side view of head; f, head from above; g, head of young larva from above; h, eggs—all enlarged (original).

deposited their eggs freely and immediately in this substance.[1] At the same time the specimens were confined in a glass dish 7.5 inches in diameter and 3 inches in height. In this dish was a layer of moist sand, covered with a layer of fresh horse manure, and the vessel was covered with a piece of gauze. On the following morning all the flies, twenty-four in number, were dead, and not a single egg had been laid. A fresh supply of flies was introduced into the same vessel, and the next morning all were dead and no eggs had been laid. The cover was now removed from this vessel and the latter placed in a glass cylinder 14 inches high, the

---

[1] The experiments which follow were conducted by Mr. D. W. Coquillett.

top of which was covered with gauze, and twenty flies introduced. This was at noon; by 4 o'clock in the afternoon no eggs could be found, but at 9 o'clock the next morning two clusters of eggshells, one cluster containing 26 and the other 45 eggs, were found. The eggs had been deposited in small cavities between the sides of the vessel and the manure, at a depth of about a quarter of an inch below the surface, but were not arranged in any regular order. Afterwards several black-berries, cherries, and partly decayed apples were placed in this vessel, and more flies were introduced. A single egg was found the next day on the upper side of one of the blackberries. At a later date experi-ments were tried in the same jar with fresh cow manure. Apparently no eggs were deposited until the third day, when two small clusters were observed. These hatched in due time, but all the larvæ died before attaining full growth.

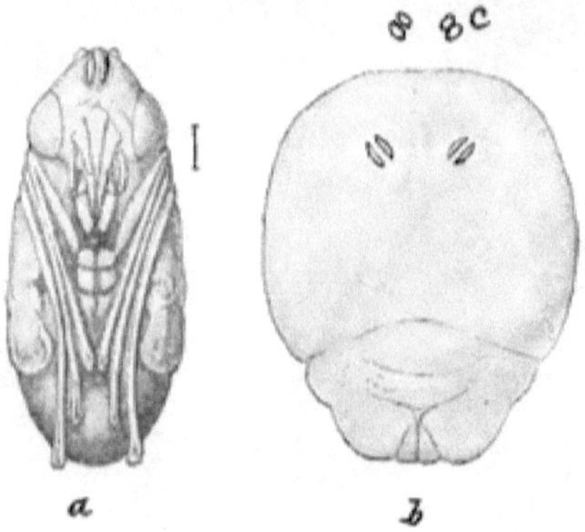

These experiments were hardly extensive enough to al-low us to generalize, but so far as they go they seem to show that horse manure is the favorite breeding place of the house fly. Continuous observations made upon the offspring of flies which bred most freely in this last-named substance indicated that the larvæ molt twice and that there are thus three distinct larval stages. The periods of development were found to be about as follows: Egg from de-position to hatching, one-third of a day; hatching of larva to first molt, one day; first to second molt, one day; second molt to pupation, three days; pupation to issuing of the adult, five days; total life round, approximately ten days. There is thus abundance of time for the development of twelve or thirteen generations in the climate of Washington every summer.

Fig. 15.—*Musca domestica*; a, pupa removed from puparium; b, hind end of body of larva in second stage; c, anal spiracles of larva in first stage—all enlarged (original).

The number of eggs laid by an individual fly is undoubtedly very large, averaging about 120, and the enormous numbers in which the insects occur is thus plainly accounted for, especially when we consider the abundance and universal occurrence of appropriate larval food. The different stages of the insect are well illustrated in the accompany-ing figures and need no description.

Taschenberg in his Praktische Insektenkunde, iv, 1880, 102–107, gives a good popular account of the house fly, but leaves the impression that the duration of a generation is much longer than we have indicated. He also states that the female lays its eggs on a great variety of sub-stances, particularly on spoiled and moist food stuffs, decaying meat,

meat broth, cut melons, dead animals, in manure pits, on manure heaps, and even in cuspidors and open snuff boxes. The fact remains however that horse manure forms the principal breeding place, and that in confinement we have been unable to rear it to maturity on any other substance.

There is not much that need be said about remedies for house flies. A careful screening of windows and doors during the summer months, with the supplementary use of sticky fly paper, is a method known to everyone, and there seems to be little hope in the near future of much relief by doing away with the breeding places. A single stable in which a horse is kept will supply house flies for an extended neighborhood. People living in agricultural communities will probably never be rid of the pest, but in cities, with better methods of disposal of garbage and with the lessening of the numbers of horses and horse stables consequent upon electric street railways and bicycles, and probably horseless carriages, the time may come, and before very long when window screens may be discarded. The prompt gathering of horse manure which may be treated with lime or kept in a specially prepared pit would greatly abate the fly nuisance, and city ordinances compelling horse owners to follow some such course are desirable. Absolute cleanliness, even under existing circumstances, will always result in a diminution of the numbers of the house fly, and, as will be pointed out in other cases in this bulletin, most household insects are less attracted to the premises of what is known as the old-fashioned housekeeper than to those of the other kind.

The house fly has a number of natural enemies, and, as will be pointed out in the next section of this bulletin, the common house centipede destroys it in considerable numbers; there is a small reddish mite which frequently covers its body and gradually destroys it; it is subject to the attacks of hymenopterous parasites in its larval condition, and it is destroyed by predatory beetles at the same time. The most effective enemy, however, is a fungous disease known as *Empusa muscæ*, which carries off flies in large numbers, particularly toward the close of the season. The epidemic ceases in December, and although many thousands are killed by it, the remarkable rapidity of development in the early summer months soon more than replaces the thousands thus destroyed.

L. O. H.

THE HOUSE CENTIPEDE.

(*Scutigera forceps* Raf.)

This centipede, particularly within the last ten or twelve years, has become altogether too common an object in dwelling houses in the Middle and Northern States for the peace of mind of the inmates. It is a very fragile creature, capable of very rapid movements, and elevated considerably above the surface upon which it runs by very numerous long legs. It may often be seen darting across floors with very great speed, occasionally stopping suddenly and remaining absolutely motionless, presently to resume its rapid movements, often darting directly at inmates of the house, particularly women, evidently with a desire to conceal itself beneath their dresses, and thus creating considerable consternation. The creature is not a true insect, but belongs to the Myriopoda, commonly known as centipedes or thousand-legs, and is sometimes called the "skein" centipede, from the fact that when crushed or motionless it looks, from its numerous long legs, like a mass of filaments or threads. It is a creature of the damp, and is particularly abundant in bathrooms, moist closets, and cellars, multiplying excessively also in conservatories, especially about places where pots are stored, and near heating pipes. In houses it will often be dislodged from behind furniture or be seen to run rapidly across the room, either in search of food or concealment. If examined closely its very cleanly habits may occasionally be manifested in that it may be observed to pass its long legs, one after another, through its mandibles, to remove any adhering dust. Its rather weird appearance, its peculiar manner of locomotion, and frequently its altogether too friendly way of approaching people, give it great interest, and, with its increasing abundance in the North, make it a subject of frequent inquiry. It is a Southern species, its normal habitat being in the southern tier of States and southwestward through Texas into Mexico. It has slowly spread northward, having been observed in Pennsylvania as early as 1849, and reaching New York and Massachusetts twenty or twenty-five years ago, but for many years after its first appearance in the latter States it was of rare occurrence. It is now very common throughout New York and the New England States, and extends westward well beyond the Mississippi, probably to the mountains.

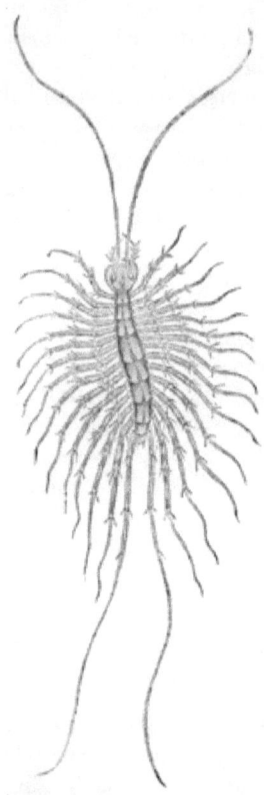

Fig. 16.—*Scutigera forceps*; Adult— natural size (original).

It is a very delicate creature, and it is almost impossible to catch it, even should one desire to do so, without dismembering several of its numerous legs or crushing it. If crushed under the foot, as one's first impulse would suggest, nothing remains but a mass of intertwined limbs, giving it the appearance of a tangle of threads. If captured, so that it can be more easily examined, it will be found to consist of a worm-like body of an inch or a little more in length, armed at the head with a pair of very long, slender antennæ, and along the sides with a fringe of fifteen pairs of long legs. The last pair are much longer than the others, in the female more than twice the length of the body. In color it is of a grayish yellow, marked above with three longitudinal dark stripes. Examination of its mouth parts shows that they are very powerful, and fitted for biting, indicating a predatory or carnivorous habit.

The indications of its mouth parts are borne out by its food habits, besides being indicated by the known food habits of the other members of the group of centipedes to which it belongs. It was inferred, before any direct observations were made, that its food was probably house flies, roaches, and any other insect inhabitants of dwellings. Later many direct observations have confirmed this inference, and in cap-

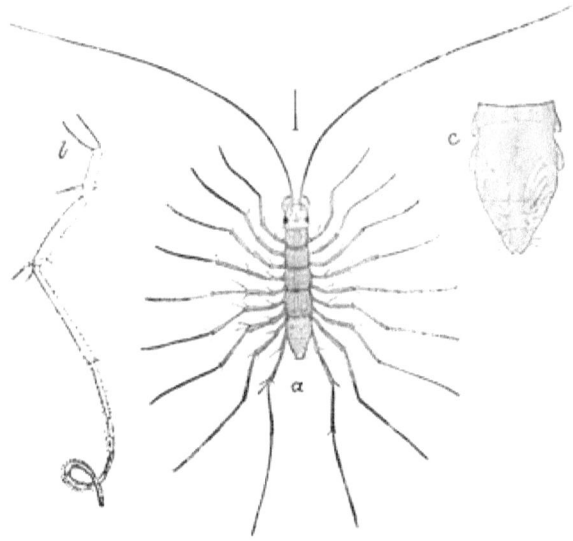

Fig. 17.—*Scutigera forceps: a*, newly-hatched individual; *b*, one of legs of same; *c*, terminal segment of body showing undeveloped legs coiled up within—all enlarged (original).

tivity, on the authority of Professor Hargitt, it feeds readily on roaches, house flies, and other insects. Miss Murtfeldt reports also having observed specimens devouring small moths. During the act of devouring a moth they kept their numerous long legs vibrating with incredible swiftness, so as to give the appearance of a hazy spot or space surrounding the fluttering moth (Insect Life, Vol. VI, p. 258). It is supposed also to feed on the bedbug, and doubtless will eat any insect which it captures, and its quickness and agility leave few insects safe from it.

Messrs. Fletcher and Howard observed its mode of capturing the croton bug, which is interesting as illustrating the habits of this centipede and its allies. In this instance the centipede sprang over its

prey, inclosing and caging it with its many legs. In its habit of springing after its prey this centipede is similar to spiders, which it also
resembles in its rapacious habits. It would therefore seem to be a very
efficient enemy of many of our house pests. The common idea that it
probably feeds on household goods and woolens or other clothing has
no basis in fact.

The popular belief is that this centipede is extremely poisonous,
and, as it belongs with the poisonous group of centipedes, it can not
be questioned but that the bite of the creature is probably somewhat
poisonous as well as painful, though the seriousness of the results will
be dependent, as in all similar cases, on the susceptibility of the patient.
The poison injected in the act of biting is probably merely to assist in
numbing and quieting its victim, and in spite of its abundance in houses
in the North, and for many years its much greater abundance in 'the
South, very few cases are recorded of its having bitten any human
being, and it is very questionable whether it would ever, unprovoked,
attack any large animal. If pressed with the bare foot or hand, or if
caught between sheets in beds, this, like almost any other insect, will
unquestionably bite in self-defense, and the few such cases on record
indicate that severe swelling and pain may result from the poison
injected. Prompt dressing of the wound with ammonia will greatly
alleviate the disagreeable symptoms.

Little is known of the early life history of this Myriapod. It is
found in the adult state in houses during practically the entire year.
Half-grown individuals are also found frequently during the summer.
A newly-born specimen was recently found by Mr. H. G. Hubbard in
the Department Insectary under a moist section of a log, and differed
from the older forms chiefly in possessing fewer legs. Its characteristics are indicated in the accompanying illustration (fig. 17). In the
half-grown and later stages it does not differ materially from the adult,
except in size, and its habits throughout life are probably subject to
little variation.

If it were not for its uncanny appearance, which is hardly calculated
to inspire confidence, especially when it is darting at one with great
speed, and the rather poisonous nature of its bite, it would not necessarily be an unwelcome visitor in houses, but, on the contrary, to be
looked upon rather as an aid in keeping in check various household
pests. Its appearance in our dwellings, however, will not often be wel-
come notwithstanding its useful rôle. It can be best controlled by
promptly destroying all the individuals which make their appearance,
and by keeping the moist places in houses free from any object behind
which it can conceal itself, or at least subjecting such locations to
freqent inspection. In places near water pipes, or in storerooms where
it may secrete itself and occur in some numbers, a free use of fresh
pyrethrum powder is to be advised.

C. L. M.

THE CLOVER MITE.

(*Bryobia pratensis* Garm.)

The subject of this section is a very minute reddish mite, less than a millimeter in length, which, particularly in the Middle States, frequently enters houses in enormous numbers in autumn, causing considerable consternation and arousing very natural fears. Aside from the

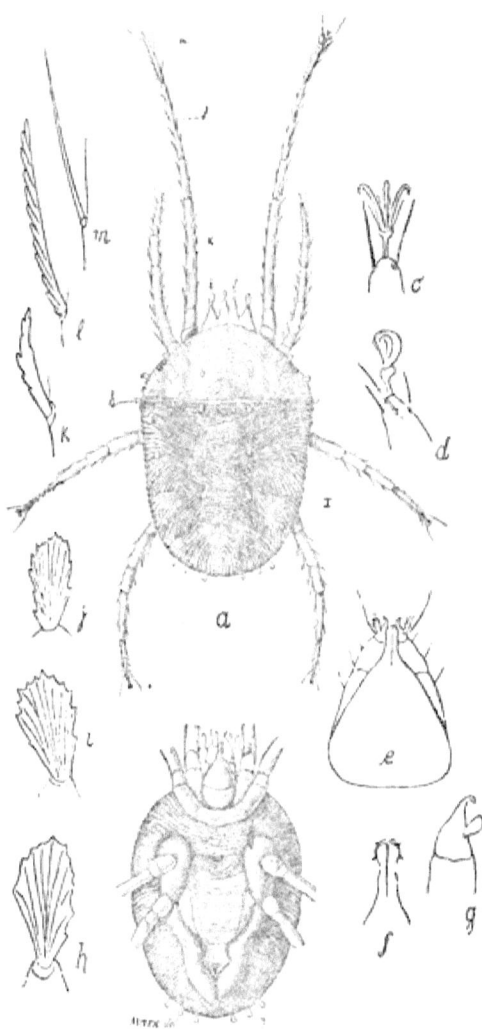

FIG. 18.—*Bryobia pratensis*: *a*, female from above; *b*, same, ventral view, with legs removed; *c* and *d*, tarsal claws; *e*, proboscis and palpi from below; *f*, proboscis enlarged; *g*, palpus enlarged; *h*, one of the body scales; *i*, scale from outer cephalo-thoracic prominence; *j*, scale from inner cephalo-thoracic prominence; *k*, serrate hair from basal joint of leg; *l*, same from penultimate joint; *m*, spine of last joint—*a*, *b*, greatly enlarged; *c-m*, still more enlarged (from Riley and Marlatt).

disagreeableness of its mere presence, it has no objectionable consequences. This mite is somewhat allied to the common red mite of
greenhouses, and in fact has a similar habit, but lives out of doors on
vegetation and has a decided preference for clover, whence its common
name of clover mite. It occurs very commonly in the Northern and
Central States from Massachusetts to California, and is frequently
abundant on various orchard and shade trees. In the mountain ranges
of the Pacific Coast its eggs have been found in enormous numbers on
the bark of various mountain trees, especially the cottonwood (*Populus
tremuloides*). These eggs are often massed two or three layers deep,

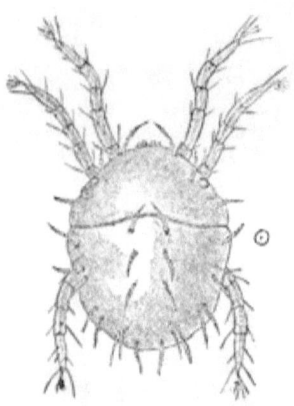

and their reddish color entirely obscures
the natural color of the bark. One writer
states that he found at least 50 square feet
of these eggs on the south sides of the
trunks of cottonwoods at an elevation of
6,000 to 8,000 feet. In the Eastern and
Central States the eggs are found similarly
placed in the crotches of orchard and shade
trees, and frequently in sufficient numbers
to give a reddish color to small areas.
Complaints of this mite have been received
from a great many sources in the Middle
and Eastern States. That they are a nuisance in houses is due to their habit of
migrating in the fall, possibly for shelter
or in search of food. In the case of house
invasions the mites will almost invariably

FIG. 19.—*Bryobia pratensis:* Newly-
hatched larva—greatly enlarged
(from Riley and Marlatt).

be found to have come from some near-by vegetation, usually from the
surrounding lawns. After they have once gained entrance they may
be exterminated by a liberal and abundant use of insect powders, fumigating with burning brimstone, or spraying with benzine, care being
taken, if the latter substance be used, to see that no fire is present. If
the invasion be discovered at the very outset, it may be stopped by
spraying the sides of the house very liberally with kerosene or by treat-
ing the surrounding lawns with a spray of kerosene emulsion.

C. L. M.

THE HOUSE CRICKET.

(*Gryllus domesticus* Linn.)

No insect inhabitants of dwellings are better known than the domestic
or hearth crickets, not so much from observation of the insects themselves as from familiarity with their vibrant, shrilling song notes,
which, while thoroughly inharmonious in themselves, are, partly from
the difficulty in locating the songster, often given a superstitious significance and taken, according to the mood of the listener, to be either

a harbinger of good and indicative of cheerfulness and plenty, or to give rise to melancholy and to betoken misfortune. The former idea prevails, however, and Cowper expresses the common belief that the—

Sounds inharmonious in themselves and harsh,
Yet heard in scenes where peace forever reigns,
And only there, please highly for their sake.

The common name "cricket" is descriptive of its cheerful, chirping note, and is derived from the imitative French popular name "cricri" (from *criquer*). Similar descriptive names are applied to it in many foreign tongues.

The introduction of the domestic cricket of Europe into America was probably at a very early date, at least in portions of the country. Kalm, a careful and scientific observer, writing in 1749 of this insect, says that they are "abundant in Canada, especially in the country, where these disagreeable guests lodge in the chimneys; nor are they uncommon in the towns. They stay here both summer and winter, and frequently cut clothes in pieces for pastime." The year before, however, he writes that he had not met with them in any of the houses in Pennsylvania or New Jersey.[1]

The occurrence of this insect in Canada in comparative abundance has since been confirmed by Provancher and Caulfield, and in various Eastern towns in the United States by Uhler, Glover, and others.

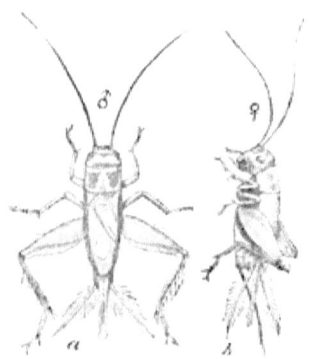

Fig. 20.—*Gryllus domesticus*: *a*, male; *b*, female—natural size (original).

It has also been observed in various States westward to and beyond the Mississippi. It does not seem to be at all common on this continent, however, except in Canada, and the more familiar insect to most Americans is one or other of our brownish-black field crickets, which often enter houses and accommodate themselves to domesticity almost as completely as the true European hearth cricket. Our native crickets are more robust and of larger size, but present the same tendency of location and food habits as their European relatives. A species (*Gryllus assimilis* Fab.) often found in houses in Washington is represented in fig. 21. The following account of the imported domestic cricket applies in the main also to any of our native species which are acquiring domesticity. Our species are, however, not known to breed in houses, although it is not at all improbable that this is now occasionally true of some of them.

The house cricket belongs to the jumping or saltatorial family of the Orthoptera, being closely allied to the common field crickets and the curious mole cricket. The normal mode of progression is by a series of leaps, the hind femora being greatly thickened and enlarged, kangaroo like. In color the house cricket is light yellowish-brown, and its squarish body and spherical head are very characteristic. The antennæ or feelers are very long and thread-like, exceeding the body in length.

The chirping song of the cricket is produced only by the male, and is supposed to be a love call. If so, it has been pointed out that it evidently betokens, on account of its long continuance, a patient persistence which deserves the highest encomium. It is produced by the friction or stridulation of the upper wings over each other. At the base of each of these wings is a large talc-like spot—the crepitaculum—which is characterized by its inflated appearance and its very coarse, irregular veining. By rasping or scraping the file-like under surface of one wing over the roughening of the other the vibrant note of the cricket is produced. The song is, therefore, analogous to that made by an instrument rather than to the voice or sounds of higher animals. To be at all significant to the insect, however, it must be heard, and what seems to be the insect ear is found in curious organs on the fore tibiæ, represented in the illustration (fig. 21, c, d, e, f).

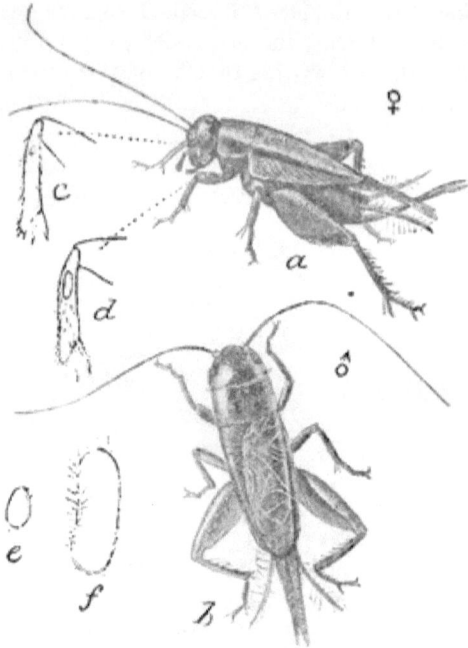

FIG. 21.—*Gryllus assimilis*: *c*, female; *b*, male; *c*, *d*, fore tibiæ, inner and outer views showing drums of ear; *e* and *f*, drums or tympana—enlarged (original).

The house cricket usually occurs on the ground floor of dwellings, and evinces its liking for warmth by often occurring in the vicinity of fireplaces, concealing itself between the bricks of chimneys or behind baseboards, frequently burrowing into the mortar of walls. It is particularly apt to abound in bakehouses. It is rarely very abundant, but at times multiplies excessively and becomes a very serious nuisance. During cold weather, or in cold rooms in winter, it remains torpid, but under the influence of warmth it becomes active and musical. It is easily kept in captivity as a pet, and will reward the possessor by furnishing an abundance of its peculiar melody, and in Spain it is often kept, it is reported, in cages, as we do singing birds. It is in the main nocturnal in its habits, coming out in the dusk of evening and roaming about the house for whatever food materials it may discover. It feeds readily on bread crumbs or almost any food product to which it can get access, and is particularly attracted to liquids, in its eagerness to get at which it often meets death by drowning. It is a very pugnacious insect and will bite vigorously if captured, and is also predaceous or carnivorous, like most of its outdoor allies. It is supposed to feed on various other house insects, such as the cockroach and is also probably cannibalistic. A pair of a native species kept in a cage by the writer, for a short period manifested the greatest friendliness, but the male shortly afterwards made a very substantial meal of his companion.

The crickets, in common with most other Orthoptera, will occasionally, in pure wantonness seemingly, cut and injure fabrics, and are particularly apt to cut into wet clothing, evidently from their liking for moisture. Any of the common field grasshoppers or crickets, entering houses, are apt to try their sharp jaws on curtains, garments, etc., and Dr. J. A. Lintner records[1] the case of a suit of clothing just from the tailor which was completely ruined in a night by a common black field cricket (*Gryllus luctuosus*), which had entered an open window in some numbers. There is a popular superstition also to the effect

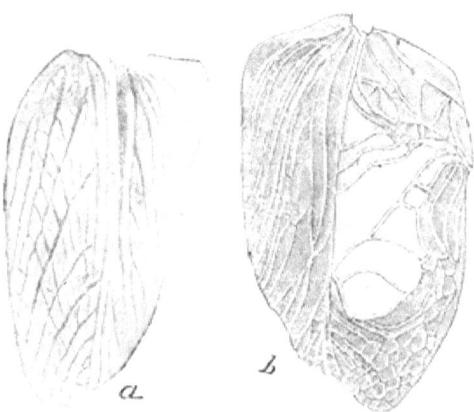

FIG. 22.—*Gryllus assimilis*: *a*, wing of female; *b*, wing of male showing more irregular and coarser veining—enlarged (original).

that if a cricket be killed its relatives will promptly cut the garments of the offender.

In Europe, and undoubtedly also in this country, the hearth cricket is found in houses in all sizes, from the very young to the full-grown insects, and probably often deposits its eggs and goes through its entire transformations within the four walls of dwellings. In summer it also appears frequently out of doors in Europe about hedges and in gardens, returning to the house for protection at the approach of cold weather, and being apparently unable to winter out of doors, at least in cold climates. In this country it has been taken at electric lights out of doors. Its eggs, judging from our knowledge of allied species, are deposited in clusters, and the young resemble their parents very closely, except in size and in lacking wings; they present also no variation in habit.

So much superstition and popular interest attaches to the house

[1] 8th Rept. Ins. N. Y., p. 179.

cricket that frequently there is a strong feeling against destroying it; and to many it is a pleasant incentive to revery, filling the mind with pleasant contemplations, and perhaps lulling the wakeful to restful sleep. Not to all, however, does it appeal in this way. and for those to whom its notes are rasping and irritating, and who fear for the safety of their garments, or are otherwise evilly disposed toward it, the following methods of control will be of interest:

It may be readily destroyed by taking advantage of its liking for liquids, and any vessel containing beer or other liquid placed about will usually result in crickets being collected and drowned in numbers. It may also be destroyed by the distribution of uncooked vegetables, such as ground-up carrots or potatoes. strongly poisoned with arsenic. In the use of poisoned baits in dwellings great care, however, should always be exercised.

C. L. M.

THE PAPER WASP.

(Vespa germanica Fab.)

It frequently happens. more particularly in suburban places and in the country, that the common yellow jackets or paper wasps. notably *Vespa germanica* Fab., will have their nests near dwellings and multiply to such an extent as to become serious nuisances about houses. to which they are attracted by the moisture about wells or to fruit refuse. Under these circumstances they become a source of some danger from the liability of their stinging horses. Unless houses are carefully screened they will frequently be attracted into them in considerable numbers, and on account of their pugnacious disposition render meal taking a proceeding of considerable risk. They have a great fondness for all sweetened liquids and will swarm over fruit. especially melons.

The species most apt to be annoying in houses in the East is the one mentioned at the head of this article. It is of European origin, and, like many other introduced animals, as the English sparrow. for example, has become even more numerous in its new home than in its old. It sometimes nests in trees in Europe, but in this country commonly dwells in large underground colonies located usually only a few inches below the surface, and often in the deserted nests of field mice, which have been cleaned out and greatly enlarged by their insect tenants.

The nest consists of a loose papery envelope. within which are from four to eight stories or tiers of combs, attached to each other with strong central supports. The largest combs sometimes have a diameter of 12 inches and the larger nests a capacity of upward of one-half bushel. Throughout the summer a colony contains, in addition to the queen mother, workers only. The perfectly sexed individuals, females or queens and males, appear only in the fall, usually in September, are much larger than the workers, and are reared in special cells of large size in the undermost or last constructed of the combs.

With the approach of cold weather the nests are abandoned, most of the individuals, including all the workers and males, perishing, and only the perfect females, the product of the last fall brood, wintering over. Early in spring these over-wintered females come out of the cracks in logs or holes in walls, etc., in which they have hibernated, and unaided originate new colonies of workers, which by midsummer often contain 20,000 or more individuals. No honey, wax, or pollen is stored in the nests, but the young are fed by the workers on a liquid derived from insects or other substances eaten.

The paper wasps have a number of natural enemies. They are captured and devoured by two species of robber flies, and in addition their underground nests, as I am informed by woodmen, are frequently dug out by foxes and skunks, which feed on the larvæ and pupæ contained in them.

The best means of abating the wasp nuisance is to discover the nest and destroy the inmates. Ordinarily by watching individual wasps the nest can be located, and the introduction of a few spoonfuls of chloroform or bisulphide of carbon into the entrance, after all have come in for the night, will suffice to destroy the inhabitants.

Other Vespas, especially the common bald-faced hornet (*Vespa maculata* Linn.), which builds large paper nests in trees, also enter houses, but not so abundantly as the small yellow and black species referred to. The slender yellowish-brown wasps (*Polistes* spp.), which build uncovered combs attached to rafters and in trees, are also frequent visitors in houses, but are not so pugnacious and will rarely attack anyone unless they are accidentally taken hold of or their nests disturbed. All of these wasps are of more or less service to housekeepers in that they are active enemies of the common house fly.

C. L. M.

CHAPTER IV

SPECIES INJURIOUS TO WOOLEN GOODS, CLOTHING, CARPETS, UPHOLSTERY, ETC.

By L. O. HOWARD and C. L. MARLATT.

THE CARPET BEETLE, OR "BUFFALO MOTH."

(*Anthrenus scrophulariæ* Linn.)

All the year round, in well heated houses, but more frequently in summer and fall, an active brown larva a quarter of an inch or less in length and clothed with stiff brown hairs, which are longer around the sides and still longer at the ends than on the back, feeds upon carpets and woolen goods, working in a hidden manner from the under surface, sometimes making irregular holes, but more frequently following the line of a floor crack and cutting long slits in a carpet.

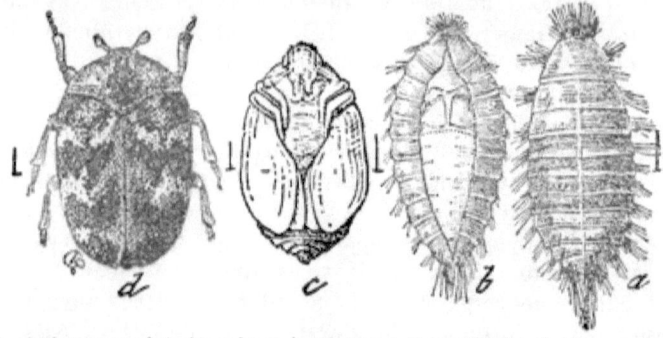

FIG. 23.—*Anthrenus scrophulariæ*: *a*, larva, dorsal view; *b*, pupa within larval skin; *c*, pupa, ventral view; *d*, adult—all enlarged—(from Riley).

This insect in the United States is known as a carpet beetle in the northern part of the country only. Beginning with Massachusetts, it extends west to Kansas. It is not known as a carpet beetle in Washington or Baltimore, and is not common in Philadelphia, but abounds in New York, Boston, all the New England States, and west through Ohio, Indiana, Michigan, Wisconsin, Illinois, Iowa, and Kansas. It is originally a European insect and is found in all parts of Europe. It was imported into this country about 1874, probably almost simultaneously at New York and Boston. It has long been known on the Pacific Coast, but not, so far as we are aware, in the rôle of a carpet enemy.

The adult insect is a small, broad-oval beetle, about three-sixteenths of an inch long, black in color, but is covered with exceedingly minute scales, which give it a marbled black-and-white appearance. It also

58

has a red stripe down the middle of the back, widening into projections at three intervals. When disturbed it "plays 'possum," folding up its legs and antennæ and feigning death. As a general thing the beetles begin to appear in the fall, and continue to issue, in heated houses, throughout the winter and following spring. Soon after issuing they pair, and the females lay their eggs in convenient spots. The eggs hatch, under favorable conditions, in a few days, and the larvæ, with plenty of food, develop quite rapidly. Their development is retarded by cold weather or lack of food, and they remain alive in the larval state, in such conditions, and particularly in a dry atmosphere, for an almost indefinite period, molting frequently and feeding upon their cast skins. Under normal conditions, however, the skin is cast about six times, and there is, probably, in the North, not more than two annual generations. When the larva reaches full growth the yellowish pupa is formed within the last larval skin. Eventually this skin splits down the back and reveals the pupa, from which the beetle emerges later. The beetles are day fliers, and when not engaged in egg laying are attracted to the light. They fly to the windows, and may often be found upon the sills or panes. Where they can fly out through an open window they do so, and are strongly attracted to the flowers of certain plants, particularly the family Scrophulariaceæ, but also to certain Compositæ, such as milfoil (*Achillea millefolium*). The flowers of Spiræa are also strongly attractive to the beetles. It is probable, however, that this migration from the house takes place, under ordinary circumstances, after the eggs have been laid.

In Europe the insect is not especially noted as a household pest, and we are inclined to think that this is owing to the fact that carpets are little used. In fact, we believe that only where carpets are extensively used are the conditions favorable for the great increase of the insect. Carpets once put down are seldom taken up for a year, and in the meantime the insect develops uninterruptedly. Where polished floors and rugs are used the rugs are often taken up and beaten, and in the same way woolens and furs are never allowed to remain undisturbed for an entire year. It is a well-known fact that the carpet habit is a bad one from other points of view, and there is little doubt that if carpets were more generally discarded in our more Northern States the "buffalo bug" would gradually cease to be the household pest that it is to-day. The insect is known in Europe as a museum pest, but has not acquired this habit to any great extent in this country. It is known to have this habit in Cambridge, Mass., and Detroit, Mich., as well as in San Francisco, Cal., but not in other localities. In all of these three cases it had been imported from Europe in insect collections.

REMEDIES.

There is no easy way to keep the carpet beetle in check. When it has once taken possession of a house nothing but the most thorough and long-continued measures will eradicate it. The practice of annual house cleaning, so often carelessly and hurriedly performed, is, as we have shown above, peculiarly favorable to the development of the insect. Two house cleanings would be better than one, and if but one, it would be better to undertake it in midsummer than at any other time of the year. Where convenience or conservatism demands an adherence to the old custom, however, we have simply to insist upon extreme thoroughness and a slight variation in the customary methods. The rooms should be attended to one or two at a time. The carpets should be taken up, thoroughly beaten, and sprayed out of doors with benzine, and allowed to air for several hours. The rooms themselves should be thoroughly swept and dusted, the floors washed down with hot water, the cracks carefully cleaned out, and kerosene or benzine poured into the cracks and sprayed under the baseboards. The extreme inflammability of benzine, and even its vapor when confined, should be remembered and fire carefully guarded against. Where the floors are poorly constructed and the cracks are wide it will be a good idea to fill the cracks with plaster of paris in a liquid state; this will afterwards set and lessen the number of harboring places for the insect. Before relaying the carpet tarred roofing paper should be laid upon the floor, at least around the edges, but preferably over the entire surface, and when the carpet is relaid it will be well to tack it down rather lightly, so that it can be occasionally lifted at the edges and examined for the presence of the insect. Later in the season, if such an examination shows the insect to have made its appearance, a good though somewhat laborious remedy consists in laying a damp cloth smoothly over the suspected spot of the carpet and ironing it with a hot iron. The steam thus generated will pass through the carpet and kill the insects immediately beneath it.

The measures used in the care of furs, rugs, and woolen goods generally to prevent the work of this insect during the summer are practically identical with those recommended for the clothes moths, elsewhere mentioned. The most perfect and simplest is storage at a temperature of from 40 to 42° F. For the cheaper methods the reader is referred to the chapter on clothes moths.

These strenuous measures, if persisted in, are the only hope of the good housekeeper, so long as the system of heavy carpets covering the entire floor surface is adhered to. Good housekeepers are conservative people, but we expect eventually to see a more general adoption of the rug or of the square of carpet, which may at all times be readily examined and treated if found necessary. Where the floors are bad the practice of laying straw mattings under the rugs produces a sightly appearance, and, while not as cleanly as a bare floor, affords still fewer harboring places for this insect.

L. O. H.

THE BLACK CARPET BEETLE.

(*Attagenus piceus* Oliv.)

This carpet beetle occurs in general in the same situations in which the preceding species is found. The larva is an active, light-brown, somewhat cylindrical creature, clothed with closely appressed hairs, and with a long terminal tuft of hairs at the end of the body. It is readily distinguished from the so-called "buffalo moth" by its shape and in general by its lighter color. It is not so fond of working in cracks and cutting long slits in carpets, and in general is not so dangerous a species as the other.

This insect has been a denizen of the United States certainly since 1854. It is widespread in Europe and Asia, and first attracted attention as a carpet insect in this country in 1879, when Dr. Lintner found

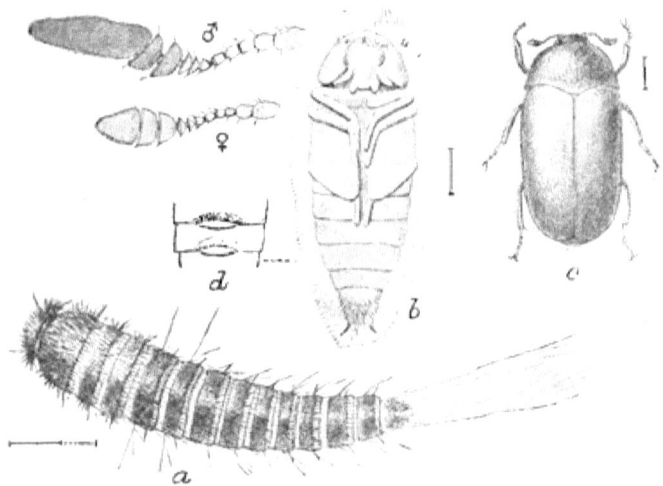

FIG. 24.—*Attagenus piceus*: *a*, larva; *b*, pupa; *c*, adult; *d*, dorsal abdominal segments of pupa; above, at left, male and female antennæ—all enlarged (original).

it in connection with the "buffalo moth" at Schenectady, N. Y. It had previously been observed by Hagen in Cambridge, in the Museum of Natural History, at an early date, and had been found in feathers by Walsh. Since 1880 it has become very abundant in Washington, D. C., and here takes the place of *Anthrenus scrophulariæ*. It has been received at the division of entomology from Goffstown, N. H.; Hartford, Conn.; New York City, Lawrence, Long Island, N. Y.; Washington and Chagrin Falls, Ohio; Detroit, Agricultural College, Charlotte, and Drain, Mich.; Philadelphia, Pa.; Wadestown, W. Va., and Memphis, Tenn. From hearsay information the writer believes that it is also more or less abundant in houses in Charleston, S. C., Savannah, Ga., and Jacksonville, Fla.

The adult insect is a small, oval, black beetle of the general appearance indicated in the figure. It is readily distinguished from *Anthrenus scrophulariae*. Its natural history has not been studied in detail, but there is little doubt that it is similar to that of the other species. It seems to have a particular predilection for feathers and has several times been observed to produce in feather beds a peculiar felting of the ticking. It has also been known to infest flour mills and is to a certain extent a feeder upon cereal products. It is a museum pest of considerable importance, and, in fact, when first discovered in connection with the Anthrenus, by Dr. Lintner, it was supposed to be present around the margin of carpets simply in search of dead flies and other animal matter, such as cast skins of Anthrenus, etc. In 1878 Dr. Hagen stated in the Proceedings of the Boston Society of Natural History that during late years this insect had propagated to a fearful extent in many houses in Cambridge, and that he believed it to be responsible for fully half of all the destruction ascribed to the previous species. In the arranged collection of the Museum of Comparative Zoology it occurred only rarely, and Dr. Hagen always found a crack or a slit in the infested box through which the thin and slender young larva had entered. The insect, he said, could always be recognized by the small, globular, ocherous excrement. Mr. Schwarz, writing in 1890, spoke of the recent increase in numbers of this insect in Washington, D. C. As a museum pest he had found it frequently in insect boxes which were not quite tight, but, fortunately, this species does not seem to be able to enter through as small a crack as Anthrenus or Trogoderma. In January, 1892, Mrs. Horace French, of Elgin, Kane County, Ill., wrote us that many houses in Elgin were infested both by this species and by the buffalo carpet beetle. The black carpet beetle, however, seemed, according to the correspondent, to work constantly through the year, unmindful of change of temperature, while the other species did little damage except during the warmer months. Her own house was completely overrun, and after taking up the carpets and discovering the full extent of their ravages it was deemed unsafe to replace them.

Until recently we had made but one attempt to follow out the detailed life history. This was in June, 1882, when the beetle seemed to be especially numerous, flying into the open windows of the office. A number were placed June 20 in a jar with pieces of rag. On June 23 six eggs were found to have been deposited, three of which were already much shriveled, apparently not fertilized. The color of the eggs was white and they were extremely soft and of broad oval shape, with irregular striate sculpturing, like the markings on the palm of one's hand. No further eggs were deposited and those previously laid did not hatch.

Quite recently, in the course of his studies of insects injurious to stored food, Mr. Chittenden, of this office, has many times met with the larva of this species in seeds and other vegetable products in the museum of the Department. He has shown that the larva will breed successfully from the egg in flour and meal. Incidentally, he observed that the beetles begin to appear in houses in Washington. D. C., as early as the last of April and occur in the greatest numbers during the hot spells late in May and early in June. By the middle of June their numbers become less. Beginning on May 6, beetles were placed from time to time into a jar with woolen cloth. On June 13 certain larvæ measuring about 1 mm. in length were found. A year from the placing of the first beetles in the jar the largest larvæ were found to be only 4.5 mm. long. Isolated full-grown larvæ were several times observed to pupate, with the result that the pupal stage was found to last from six to fifteen days. In Mr. Chittenden's experiments in rearing this insect two years were required for its development from egg to beetle.

REMEDIES.

Owing to the similarity of habits, the same remedies may be used against this insect as against the buffalo carpet beetle. Notwithstanding Mrs. French's experience to the contrary, we do not consider it as serious a household pest as the other species.

L. O. H.

THE CLOTHES MOTHS.

(Tinea pellionella, et al.)

The destructive work of the larvæ of the small moths commonly known as clothes moths, and also as carpet moths, fur moths, etc., in woolen fabrics, fur, and similar material, during the warm months of summer in the North, and in the South at any season, is an altogether too common experience. The preference they so often show for woolen or fur garments gives these insects a much more general interest than is perhaps true of any other household pest. Not only are they a pest to the

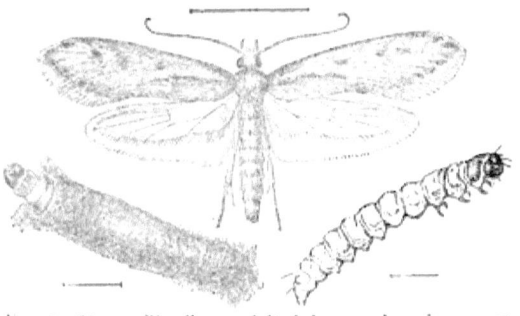

Fig. 25.—Tinea pellionella: a, adult; b, larva; c, larva in case—enlarged (from Riley).

good housekeeper, but the bachelor, whose interest in domestic matters might otherwise remain at a low ebb, knows to his sorrow of their abundance in the disastrous results of their presence in his wardrobe.

The little yellowish or buff-colored moths sometimes seen flitting about rooms, attracted to lamps at night or dislodged from infested garments, are themselves harmless enough; in fact their mouth parts are rudimentary, and they can not enjoy even the ordinary pleasures of the winged existence of other moths in sampling the nectar of flowers. It is, therefore, to the larvæ only that the destructive work is due.

The clothes moths all belong to the group of minute Lepidoptera known as Tineina, the old Latin name for cloth worms of all sorts, and are characterized by very narrow wings, fringed with long hairs. The common species of clothes moths have been associated with man from the earliest times and are thoroughly cosmopolitan. They are all probably of Old World origin, none of them being indigenous to the United States. That they were well known to the ancients is shown by Job's reference to "a garment that is moth eaten," and Pliny has given such an accurate description of one of them as to lead to the easy identifica-· tion of the species. That they were early introduced into the United States is shown by Pehr Kalm, the Swedish scientist whom we have previously quoted and who seemed to take a keen interest in house pests. He reported these Tineids to be abundant in 1748 in Philadelphia, then a straggling village, and says that clothes, worsted gloves, and other woolen stuffs hung up all summer were often eaten through and through by the worms, and furs were so ruined that the hair would come off in handfuls.[1]

What led to the first association of these and other household pests with man is an interesting problem. In the case of the clothes moths, the larvæ of all of which can, in case of necessity, still subsist on almost any dry animal matter, their early association with man was probably in the rôle of scavengers, and in prehistoric times they probably fed on waste animal material about human habitations and on fur garments. The fondness they exhibit nowadays for tailor-made suits and expensive products of the loom is simply an illustration of their ability to keep pace with man in his development in the matter of clothing from the skin garments of savagery to the artistic products of the modern tailor and dressmaker.

Three common destructive species of clothes moths occur in this country. Much confusion, however, exists in all the early writings on these insects, all three species being inextricably mixed in the descriptions and accounts of habits. Collections of these moths were submitted some years ago by Professors Fernald and Riley to Lord Walsingham, of Merton Hall, England, the world's authority on Tineids, and from the latter's careful diagnosis it is now possible to easily separate and recognize the diffent species.

The common injurious clothes moths are the case-making species (*Tinea pellionella* Linn.), the webbing species or Southern clothes moth (*Tineola biselliella* Hummel), and the gallery species or tapestry moth (*Trichophaga tapetzella* Linn.).

A few other species which normally infest animal products may

---

[1] Kalm's Travels, Vol. I, p. 317.

occasionally also injure woolens, but are not of sufficient importance to be here noted.

The case-making clothes moth (*Tinea pellionella* Linn.) (fig. 25) is the only species which constructs for its protection a true transportable case. It was characterized by Linnæus and carefully studied by Réaumur early in the last century. Its more interesting habits have caused it to be often a subject of investigation, and its life history will serve to illustrate the habits of all the clothes moths.

The moth expands about half an inch, or from 10 to 14 mm. Its head and forewings are grayish yellow, with indistinct fuscous spots on the middle of the wings. The hind wings are white or grayish and silky. It is the common species in the North, being widely distributed and very destructive. Its larva feeds on woolens, carpets, etc., and is especially destructive to furs and feathers. In the North it has but one annual generation, the moths appearing from June to August, and, on the authority of Professor Fernald, even in rooms kept uniformly heated night and day it never occurs in the larval state in winter. In the South, however, it appears from January to October, and has two or even more broods annually.

Pliny says of its larva that it "is clad in a jacket, gradually forming for itself its own garment, like the snail in its shell, and when this is taken from it, it immediately dies; but when its garment has reached its proper dimensions it changes into a chrysalis, from which, at the proper time, the moth issues."

The larva is a dull white caterpillar, with the head and the upper part of the next segment light brown, and is never seen free from its movable case or jacket, the construction of which is its first task. If it be necessary for it to change its position, the head and first segment are thrust out of the case, leaving the thoracic legs free, with which it crawls, dragging its case after it to any suitable situation. With the growth of the larva it becomes necessary from time to time to enlarge the case both in length and circumference, and this is accomplished in a very interesting way. Without leaving its case the larva makes a slit halfway down one side and inserts a triangular gore of new material. A similar insertion is made on the opposite side, and the larva reverses itself without leaving the case and makes corresponding slits and additions in the other half. The case is lengthened by successive additions to either end. Exteriorly the case appears to be a matted mass of small particles of wool; interiorly it is lined with soft, whitish silk. By transferring the larva from time to time to fabrics of different colors the case may be made to assume as varied a pattern as the experimenter desires, and will illustrate, in its coloring, the peculiar method of making the enlargements and additions described.

On reaching full growth the larva attaches its case by silken threads to the garment or other material upon which it has been feeding, or sometimes carries it long distances. In one instance numbers of them were noticed to have scaled a 15-foot wall to attach their cases in an angle of the cornice of the ceiling. It undergoes its transformations to the chrysalis within the larval case, and under normal conditions the moth emerges three weeks later, the chrysalis having previously worked partly out of the larval case to facilitate the escape of the moth. The latter has an irregular flight and can also run rapidly. It has a distinct aversion to light and usually promptly conceals itself in garments or crevices whenever it is frightened from its resting place. The moths are comparatively short lived, not long surviving the deposition of their eggs for a new generation of destructive larvæ. The eggs are minute, not easily visible to the naked eye, and are commonly placed directly on the material which is to furnish the larvæ with food. In some cases they may be deposited in the crevices of trunks or boxes, through which the newly hatched larvæ enter.

In working in feathers this insect occasionally causes a felting very similar to that produced by the dermestid beetle *Attagenus piceus* (p. 61).

The protection afforded by the seclusion of this insect in houses does not prevent its having insect enemies, and at least two hymenopterous parasites have been reared in this country from its larval cases. These are *Hyperaemus tineæ* Riley MS., and *Apanteles carpatus* Say, both reared from specimens collected in Michigan.

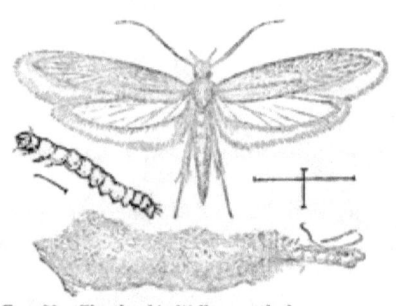

FIG. 26.—*Tineola biselliella*: moth, larva, cocoon and empty pupa-skin—enlarged (after Riley)

The webbing or Southern clothes moth (*Tineola biselliella* Hummel) (fig. 26) is the more abundant and injurious species in the latitude of Washington and southward. It also occurs farther north, though in somewhat less numbers than the preceding species. It presents two annual broods even in the Northern States, the first appearing in June from eggs deposited in May, and the second in August and September. It is about the size of *pellionella*. The forewings are, however, uniformly pale ocherous, without markings or spots. Its larva feeds on a large variety of animal substances—woolens, hair, feathers, furs, and in England it has even been observed to feed on cobwebs in the corners of rooms, and in confinement has been successfully reared on this rather dainty food substance. The report that it feeds on dried plants in herbaria is rather open to question, as its other recorded food materials are all of animal origin.

Frequently this species is a very troublesome pest in museums, particularly in collections of the larger moths. Prof. F. M. Webster, of Wooster, Ohio, has had some of his large moths badly riddled by its larvæ, and Dr. Hagen also records it as feeding on insect collections.

Dr. Riley reared it in conjunction with the angoumois grain moth (*Sitotroga cerealella*) from grain, it being apparent that its larvæ had subsisted on dead specimens of the grain moth. It is very apt to attack large Lepidoptera on the spreading board, and has, in fact, been carried through several generations on dried specimens of moths.

Its general animal-feeding habit is further indicated by the interesting case reported by Dr. J. C. Merrill, U. S. A., who submitted a sample can of beef meal which had been rejected as "weevilly." The damage proved to be due to the larvæ of *Tineola biselliella* and goes to substantiate the theory already advanced that clothes moths were scavengers in their earliest association with man.

The larva of this moth constructs no case, but spins a silky or more properly cobwebby path wherever it goes. When full grown it builds a cocoon of silk, intermixed with bits of wool, resembling somewhat the case of *pellionella*, but more irregular in outline. Within this it undergoes its transformation to the chrysalis, and the moth in emerging leaves its pupal shell projecting out of the cocoon, as with the preceding species.

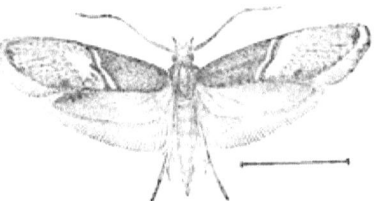

Fig. 27.—*Trichophaga tapetzella*; adult moth—enlarged (after Riley).

The tapestry moth (*Trichophaga tapetzella* Linn.) (fig. 27) is rare in the United States. It is much larger than either of the other two species, measuring three-fourths inch in expanse of wings, and is more striking in coloration. The head is white, the basal third of the forewings black, with the exterior two thirds of a creamy white, more or less obscured on the middle with gray; the hind wings are pale gray.

It normally affects rather coarser and heavier cloths than the smaller species and is more apt to occur in carpets, horse blankets, and tapestries than in the finer and thinner woolen fabrics. It also affects felting, furs, and skins, and is a common source of damage to the woolen upholstering of carriages, being rather more apt to occur in carriage houses and barns than in dwelling houses. Its larva enters directly into the material which it infests, constructing burrows or galleries lined more or less completely with silk. Within these galleries it is protected and concealed during its larval life, and later undergoes its transformations without other protection than that afforded by the gallery. The damage is due as much or more to its burrowing as to the actual amount of the material consumed for food.

One of the parasites reared from *pellionella* (*Apanteles carpatus* Say) has also been reared from this species at St. Louis, Mo.

REMEDIES.

There is no easy method of preventing the damage done by clothes moths, and to maintain the integrity of woolens or other materials which they are likely to attack demands constant vigilance, with frequent inspection and treatment. In general they are liable to affect injuriously only articles which are put away and left undisturbed for some little time. Articles in daily or weekly use, and apartments frequently aired and swept, or used as living rooms, are not apt to be seriously affected. Carpets under these conditions are rarely attacked, except sometimes around the borders, where the insects are not so much disturbed by walking and sweeping. Agitation, such as beating, shaking or brushing, and exposure to air and sunlight are old remedies and still among the best at command. Various repellants, such as tobacco, camphor, naphthaline cones or balls, and cedar chips or sprigs, have a certain value if the garments are not already stocked with eggs or larvæ. The odors of these repellants are so disagreeable to the parent moths that they are not apt to come to deposit their eggs as long as the odor is strong. As it weakens the protection decreases, and if the eggs or larvæ are already present, these odors have no effect on their development; while if the moths are inclosed with the stored material to be protected by these repellants, so that they can not escape, they will of necessity deposit their eggs and the destructive work of the larvæ will be little, if at all, restricted. After woolens have been given a vigorous and thorough treatment and aired and exposed to sunlight, however, it is of some advantage in packing them away to inclose with them any of the repellants mentioned. Cedar chests and wardrobes are of value in proportion to the freedom of the material from infestation when stored away; but as the odor of the wood is largely lost with age, in the course of a few years the protection greatly decreases. Furs and garments may also be stored in boxes or trunks which have been lined with the heavy tar paper used in buildings. New papering should be given to such receptacles every year or two. Similarly, the tarred-paper moth bags are of some value, always, however, first subjecting the materials to the treatment outlined above.

To protect carpets, clothes, and cloth-covered furniture, furs, etc., they should be thoroughly beaten, shaken, brushed, and exposed as long as is practicable to the sunlight in early spring, either in April, May, or June, depending on the latitude. The brushing of garments is a very important consideration, to remove the eggs or young larvæ, which might escape notice. Such material can then be hung away in clothes closets which have been thoroughly cleaned and, if necessary, sprayed with benzine about the cracks of the floor and the baseboards. If no other protection be given, they should be examined at least once a month during summer, brushed, and, if necessary, exposed to the sunlight.

It would be more convenient, however, to so inclose or wrap up such material as to prevent the access of the moths to it, after it has once been thoroughly treated and aired. This can be easily effected in the case of clothing and furs by wrapping them up tightly in stout paper, or inclosing in well-made bags of cotton or linen cloth or strong paper. Dr. Howard has adopted a plan which is inexpensive and which he has found eminently satisfactory. For a small sum he secured a number of the large pasteboard boxes such as tailors use, and in these packs away all winter clothing, gumming a strip of wrapping paper around the edge, so as to seal up the box completely and leave no cracks. These boxes with care will last many years. With thorough preliminary treatment it will not be necessary to use the tar-impregnated paper sacks sold as moth protectors, which may be objectionable on account of the odor.

The method of protection adopted by one of the leading furriers of Washington, who also has a large business and experience in storing costly furs, etc., is practically the course already outlined.

Furs, etc., when received are first most thoroughly and vigorously beaten with small sticks, to dislodge all loosened hair and the larvæ or moths. They are then gone over carefully with a steel comb and packed away in large boxes lined with heavy tar roofing paper, or in closets similarly lined with this paper. An examination is made every two to four weeks, and, if necessary at any time, any garment requiring it is rebeaten and combed. During many years of experience in this climate, which is especially favorable to moth damage, this merchant has prevented any serious injury from moths.

A common method of protection followed by larger dealers in carpets and furs, etc., is the use of cold storage for protection. In all large towns anyone can avail himself of this means by patronizing storage companies, and protection will be guaranteed. A temperature maintained at 40° F. is protective, but often a much lower temperature is maintained—down to 20° F.

In the case of cloth-covered furniture and cloth-lined carriages which are stored or left unused for considerable periods in summer it will probably be necessary to spray them twice or three times, viz, in April, June, and August with benzine or naphtha, to protect them from moths. These substances can be applied very readily with any small spraying device and will not harm the material, but caution must be exercised on account of their inflammability. Another means of protecting such articles is to sponge them very carefully with a dilute solution of corrosive sublimate in alcohol made just strong enough not to leave a white stain.

C. L. M.

CHAPTER V.

SPECIES INJURIOUS TO WALL PAPER, BOOKS, TIMBERS, ETC.

By C. L. MARLATT.

THE WHITE ANT.

(*Termes flaripes* Koll.)

No insect occurring in houses is capable of doing greater damage than the one under consideration. Its injuries are often hidden and concealed until the damage is beyond repair, and as it affects the integrity of the building itself as well as its contents, the importance of the insect becomes very evident. Fortunately it is not often present in the North in houses, but as the Tropics are approached the injuries from it in dwellings or other structures of wood are of common experience and often of the most serious nature, causing the sudden crumbling of bridges, wharves, and settling of floors or buildings.

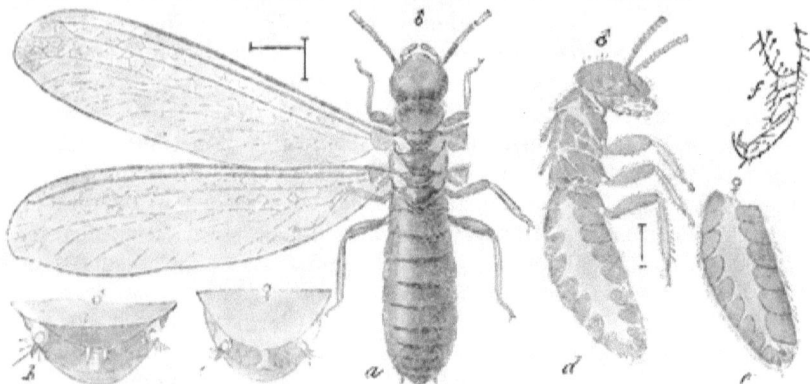

FIG. 28.—*Termes flaripes: a*. adult male; *b*. terminal abdominal segments of same from below; *c*. same of female; *d*. male. side view somewhat inflated by treatment with ammonia; *e*. abdomen of female. side view; *f*, tarsus, showing joints and claw; *a, d, e*. enlarged; *b, e, f*, greatly enlarged (original).

The term "white ant," by which this insect is universally known, is entirely inappropriate in so far as it indicates any relationship with the true ants. Strictly speaking, the white ant is not an ant at all, but belongs with the Neuroptera and is allied to the dragon flies and May flies. The only analogy with ants is in superficial resemblance and in the social habits of the two groups, in which great similarity exists. The popular acquaintance with the termite or white ant is mainly derived from witnessing its nuptial spring flight, when the small, brownish, ant-like creatures with long glistening white wings emerge from cracks in the ground or from crevices in buildings, swarming out sometimes in enormous numbers, so that they may often be swept up by the quart. These winged individuals are not the ones which do the damage, however, and are a mere colonizing form. The real depredators are soft-bodied, large-headed, milky-white insects, less than a quarter of an inch in length, which may often be found in numbers under rotting boards or in decaying stumps. These last are the workers and soldiers (fig. 31, *c* and *d*), and constitute the bulk of the colony for most of the year, the winged migrating forms, consisting of the sexed individuals, appearing normally only once a year, usually in spring or early summer.

The white ants present, in an entirely distinct order of insects, another of those most curious problems of communal societies which find so many examples among the ants, bees, and wasps. A colony of white ants includes workers, soldiers, the young of the various forms, and, at the proper season of the year, the winged males and females; also a single parent pair, the specially developed king and queen. In the case of the common white ant of this country (*Termes flavipes*), the true fully developed queen or mother of the colony and her consort, the fully developed king or male, have never been found. The soldiers or workers are degraded or undeveloped individuals of both sexes, differing in this respect from ants and bees, in which the workers are all undeveloped females.

The economy of the termites is almost exactly analogous to that of the ants and bees. The workers attend to all the duties of the colony, make the excavations, build the nests, care for the young, and protect and minister to the wants of the queen or mother ant. In this they are assisted somewhat by the soldiers, whose duty, however, is also protective, their enormous development of head and jaws indicating their rôle as the fighters or defenders of the colony. Both the workers and soldiers are blind. The colonizing individuals differ from the others in being fully developed sexually and in possession of very long wings, which normally lie flat over each other, the upper wings concealing the lower, and both projecting beyond the abdomen. These wings have a very peculiar suture near the base, where they can be readily broken off, leaving mere stumps. At the time of the spring flight the winged individuals emerge from the colony very rapidly, frequently swarming in clouds out of doors, and after a short flight fall to the ground and very soon succeed in breaking off their long, clumsy wings at the suture referred to. In this swarming or nuptial flight they come out in pairs and under favorable conditions each pair might establish a new colony, but in point of fact this probably rarely if ever happens. They are weak flyers, clumsy, and not capable of extensive locomotion on foot, and are promptly preyed upon and destroyed by many insectivorous animals, and rarely indeed do any of the individuals escape.

Theoretically, if one of these pairs succeeded in finding a decaying stump or other suitable condition at hand, they would enter it, and the king and queen, being both active, would attend to the wants of the new colony and superintend the rearing of the first brood of workers and soldiers, which would then assume the laborious duties of the young colony. Thereafter the queen, by constant and liberal feeding and absolute inaction, would increase immensely, her abdomen becoming many thousand times its original size. She would practically lose the power of locomotion and become a mere egg-laying machine of enormous capacity. Allied species whose habits have been studied in this particular indicate an egg-laying rate of 60 per minute, or something like 80,000 per day.

In the absence of a queen, however, white ants are able to develop from a very young larva or a nymph of what would otherwise become a winged female what is known as a supplementary queen, which is never winged and never leaves the colony. This supplementary queen (fig. 31, *a*) is smaller than the perfect sexed queen, but subserves all the needs

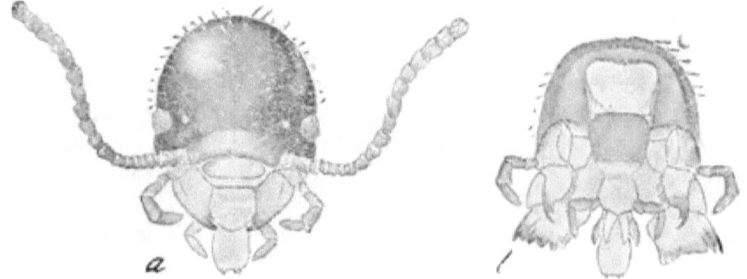

Fig. 29.—*Termes flavipes;* *a*, head of winged female viewed from above; *b*, same from below, with mouth parts opened out —greatly enlarged (original).

of the colony in the matter of egg laying, and is the only parent insect so far found in the nests of the common white ant in this country. Whether a true queen exists or not is, therefore, open to question; if not, all the individuals which escape in the spring and summer migrations must perish, and this swarming would, therefore, have to be considered a mere survival of a once useful feature in the economy of this insect, now no longer, or rarely, of service.

The normal method of the formation of new colonies is probably by the mere division or splitting up of old ones or the carrying of infested logs or timbers from one point to another.

The development of these curious insects is very simple. There is scarcely any metamorphosis, the change from the young larva to the adult being very gradual and without any marked difference in structure. They feed on decaying wood or vegetable material of any sort, and are able to carry their excavations into any timbers which are moistened, or into furniture, books, or papers stored in rooms which are at all moist. Their food is the finely divided material into which they bore, and from which they seem to be able to extract a certain amount of nourishment, sometimes redevouring the same material several times. They are also somewhat cannibalistic, and will devour the superfluous members of the colony without compunction, and normally consume all dead individuals, cast skins, and other refuse material. They may also feed to a certain extent on the liquids produced by the decaying vegetable matter in which they live, and perhaps on the fungoid elements resulting from such decay. They are capable also of exuding a sort of nectar, which is used to feed the young and the royal pair, and which they also generously give to each other.

All except the migrating winged forms are incapable of enduring full sunlight, and the soft, delicate bodies of the workers, soldiers, and young rapidly shrivel when exposed. In all their operations, therefore, they carefully conceal themselves, and in their mining of timbers or books and papers the surface is always left intact, and whenever it is necessary for them to extend their colonies it is only done under the protection of covered runways, which they construct of particles of comminuted wood or little pellets of excrement. In this way the damage which they are doing is often entirely hidden, and not until furniture breaks down or the underpinning and timbers of houses or floors yield

FIG. 30.—*Termes flavipes*: a, newly-hatched larva; b, same from below; c, egg—all enlarged to same scale (original).

is the injury recognized. The swarming of winged individuals in the early summer, if in or about houses, is an indication of their injurious presence and warrants an immediate investigation to prevent serious damage later on.

The common termite of America is very widespread, occurring from the Atlantic to the Pacific and from Canada southward to the Gulf. It has been found on the mountains of Colorado and Washington at a height of over 7,000 feet. In prairie regions it may often be seen during the swarming season issuing from the ground at frequent intervals over large pasture tracts, where it must feed on the roots of grass and other herbage. It has also been carried to other countries and is a common and often very injurious enemy of buildings and libraries in Europe. A closely allied and equally injurious European species (*Termes lucifugus*) has also been brought to this country in exchange for ours, but compared with our own species is somewhat rare though already widely distributed. In this country serious damage to buildings from the white ant has not been of common occurrence, especially in the North, except in some notable instances. In Europe our species has caused greater damage, and some years ago gained access to one of the Imperial hothouses at Vienna, and in spite of all efforts to save the building it was necessary ultimately to tear it down and replace it with an iron structure. In this country instances are on record of very serious damage to books and papers. An accumulation of books and papers belonging to the State of Illinois was thoroughly ruined by their attacks. A school library in South Carolina, which had been left closed for the summer, was found on being opened in the autumn to be completely eaten out and rendered valueless. In the Department of Agriculture an accumulation of records and documents stored in a vault which was not thoroughly dry, and allowed to remain undisturbed for several years, on examination proved to be thoroughly

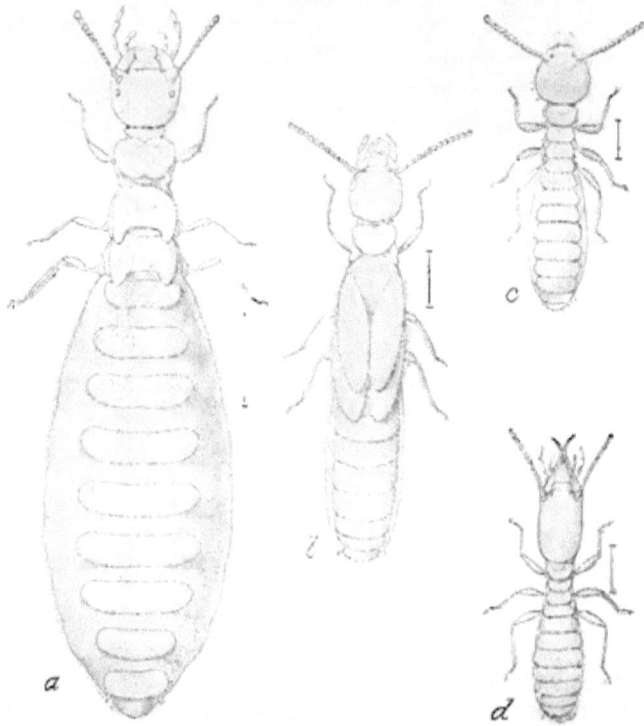

Fig. 51.—*Termes flavipes*: a, queen; b, nymph of winged female; c, worker; d, soldier—all enlarged (original).

mined and ruined by white ants. Humboldt, on the authority of Hagen, accounts for the rarity of old books in New Spain by the frequency of the destructive work of these insects.

Numerous instances of damage to underpinning of buildings and to timbers are also on record. The flooring of one of the largest sections of the United States National Museum has, for some years back, been annually undermined and weakened by a very large colony of these pests which could not be located, and finally the present season the authorities solved the problem by replacing the wood floor with one of cement. A few years ago it was found necessary to tear down and

rebuild three frame buildings in Washington in consequence of the
work of this insidious foe.

Damage of the sort mentioned has occurred as far north as Boston,
but, as stated, greatly increases as one approaches the Tropics, where the
warmth and moisture are especially suited to the development and multiplication of these insects. Here houses and furniture are never safe
from attack. The sudden crumbling into masses of dust of chairs, desks,
or other furniture, and the mining and destruction of collections of books
and papers, are matters of common experience, very little hint of the
damage being given by a surface inspection, even when the interior of
timbers or boards has been thoroughly eaten out, leaving a mere paper
shell. While confining their work almost solely to moistened or decaying timbers or vegetable material of any sort, books, and papers that
are somewhat moist, they are known to work also in living trees, carrying their mines through the moist and nearly dead heart wood. In this
way some valuable trees in Boston were so injured as to make their
removal necessary. In Florida they are often the cause of great damage
to orange trees, working around the crowns and in the roots of trees.
They are sometimes also the occasion of considerable loss in conservatories, attacking cuttings and the roots of plants. In prairie regions
also their work must necessarily be of the latter nature.

The white ant is not confined to country places, but is just as apt to
occur in the midst of towns, and especially in buildings which are surrounded by open lawns containing growing trees and flower beds richly
manured.

The first means of protection, therefore, consists in surrounding all
libraries or buildings in which articles of value are stored with clear
spaces and graveled or asphalted walks. The normal habit of these
insects of breeding in decaying stumps and partially rotted posts or
boards immediately suggests the wisdom of the prompt removal of all
such material which would otherwise facilitate the formation or per-
petuation of their colonies. Complete dryness in buildings is an important means of rendering them safe from attack, and the presence of
flying termites at any time in the spring or summer should be followed
immediately by a prompt investigation to locate the colony and determine the possibilities of damage. The point of emergence of winged
individuals may approximately, though not always, indicate the location
of the colony, and if it can be got at by the removal of flooring or
opening the walls, the colony may be destroyed by the removal of the
decaying or weakened timbers and a thorough drenching with steam,
hot water, or, preferably, kerosene or some other petroleum oil. The
destruction of winged individuals as they emerge is of no value whatever; the colony itself must be reached or future damage will not be
interfered with in the least. If the colony be inaccessible it may sometimes be possible to inject into the walls or crevices, from which the
winged individuals are emerging, kerosene in sufficient quantity to
reach the main nest, if the conditions be such as to indicate that it may

be near by, and by this means most. if not all. of the inmates may be killed.   In all districts of the South frequent examinations of libraries and stored papers should be made.

The advisability, in regions where the ant is likely to be especially destructive, of giving all buildings a stone foundation or imbedding all the lower timbers and joists in cement will be at once evident.

THE SILVER FISH.

(*Lepisma saccharina* Linn.)

This insect is often one of the most troublesome enemies of books, papers, card labels in museums, and starched clothing, and occasionally stored food substances. Its peculiar fish-like form and scaly, glistening body, together with its very rapid movements and active efforts at

concealment whenever it is uncovered, have attached considerable popular interest to it and have resulted in its receiving a number of more or less descriptive popular names, such as silver fish, silver louse, silver witch, sugar fish, etc. The species named above is the common one in England, but also occurs in this country, and, like most other domestic insects, is now practically cosmopolitan. It has a number of near allies, which closely resemble it, both in appearance and habits. One of these (*Lepisma (Thermobia) domestica* Pack.) has certain peculiarities of habit which will be referred to later. The peculiar appearance of the common silver fish early drew attention to it, and a fairly accurate description of it, given in a little work published in London in 1665 by the Royal Society, is

FIG. 32.—*Lepisma saccharina:* adult—enlarged (original).

interesting enough to reproduce:

It is a small, silvery, shining worm or moth which I found much conversant among books and papers, and is supposed to be that which corrodes and eats holes through the leaves and covers. It appears to the naked eye a small, glittering, pearl-colored moth, which, upon the removing of books and papers in the summer, is often observed very nimbly to send and pack away to some lurking cranny where it may better protect itself from any appearing dangers. Its head appears big and blunt, and its body tapers from it toward the tail, smaller and smaller, being shaped almost like a carret.[1]

On account of its always shunning the light and its ability to run very rapidly to places of concealment, it is not often seen and is most

---

[1] Micrographia, R. Hooke, London, 1665.

difficult to capture, and being clothed with smooth, glistening scales, it will slip from between the fingers and is almost impossible to secure without crushing or damaging. It is one of the most serious pests in libraries, particularly to the binding of books, and will frequently eat off the gold lettering to get at the paste beneath, or, as reported by Mr. P. R. Uhler, of Baltimore, often gnaws off white slips glued on the backs of books. Heavily glazed paper seems very attractive to this insect, and it has frequently happened that the labels in museum collections have been disfigured or destroyed by it, the glazed surface

having been entirely eaten off. In some cases books printed on heavily sized paper will have the surface of the leaves a good deal scraped, leaving only the portions covered by the ink. It will also eat any starched clothing, linen, or curtains, and has been known to do very serious damage to silks which had probably been stiffened with sizing. Its damage in houses, in addition to its injury to books, consists in causing the wall paper to scale off by its feeding on the starch paste. It occasionally gets into vegetable drugs or similar material left undisturbed for long

Fig. 33.—*Lepisma domestica*—adult female—enlarged (original).

periods. It is reported also to eat occasionally into carpets and plush-covered furniture, but this is open to question.

The silver fish belongs to the lowest order of insects—the Thysanura—is wingless, and of very simple structure. It is a worm like insect about one-third of an inch in length, tapering from near the head to the extremity of the body. The head carries two prominent antennæ, and at the tip of the body are three long, bristle-shaped appendages, one pointing directly backward and the other two extending out at a considerable angle. The entire surface of the body is covered with very minute scales like those of a moth. Six legs spring from the thorax, and, while not very long, they are powerful and enable the insect to run with great rapidity.

In certain peculiarities of structure, and also in their habits, these anomalous insects much remind one of roaches, and their quick, gliding movements and flattened bodies greatly heighten this resemblance. More striking than all, however, is the remarkable development of the coxæ or basal joints of the legs in the silver fish, which finds its counterpart in roaches, and, taken in connection with the other features of resemblance, seems to point to a very close alliance between the two groups, if, indeed, the silver fish are not merely structurally degraded forms of roaches and to be properly classed with the Blattidæ.

The general distribution of the insect about rooms, in bookcases, and under wall paper renders the application of insecticides difficult and often impracticable. It readily succumbs to pyrethrum, and whereever this can be applied, as on book shelves, it furnishes the best means of control. For starched clothing and similar objects liable to be injured by it there are no means except frequent handling and airing and the destruction by hand of all specimens discovered. Little damage is liable to occur in houses except in comparatively moist situations or where stored objects remain undisturbed for a year or more.

Another of the common silver fishes of this country, referred to in the opening paragraph, has developed a novel habit of frequenting ovens and fireplaces, and seemingly revels in an amount of heat which would be fatal to most other insects. It disports itself in numbers about the openings of ranges and over the hot bricks and metal, manifesting a most surprising immunity from the effects of high temperature. This heat-loving or bakehouse species (fig. 33) was described in 1873 as *Lepisma domestica* by Packard, who reported it to be common about fireplaces at Salem, Mass. This species is also very abundant in Washington. What is evidently this same insect has become very common, particularly in the last year or two, in England and on the Continent, where it manifests the same liking for hot places exhibited by it in this country. The habit of this species of congregating in bakehouses and dwellings, about fireplaces and ovens, has given rise to the common appellation for it in England of " fire-brat." Similar descriptive names are applied to them also on the Continent. This species closely resembles the common silver fish in size and general appearance, but may be readily distinguished from it by the presence on the upper surface of dusky markings. It also possesses well-marked structural differences, which have led to its late reference to a distinct genus— Thermobia. An Italian entomologist, Rovelli, has described this insect under the descriptive name *furnorum*, from its inhabiting ovens, and the name of the genus to which it is now assigned by English entomologists is also descriptive of its heat-loving character. A Dutch entomologist, Oudemans, reports that he has found it in abundance in all bakehouses that he has examined in Amsterdam, where it is well known to bakers and has received a number of familiar names.

THE BOOK-LOUSE.

(*Atropos divinatoria* Fab.)

This pale, louse-like insect, measuring less than 1 mm. in length, usually occurs in houses, though rarely in any numbers, and is most often seen on opening old musty volumes, scampering across the page to conceal itself elsewhere. From this habit comes its popular name of book-louse. It is one of the smallest of insects, nearly colorless, and almost invisible to the unaided eye, except as its active movements attract one's attention. It belongs to the family Psocidæ, and is somewhat closely allied to the white ants, belonging in the same order. There are a number of species of psocids which frequent houses, all popularly styled book-lice, and having habits and characteristics very similar to the one named above, which is the more common and annoying species,

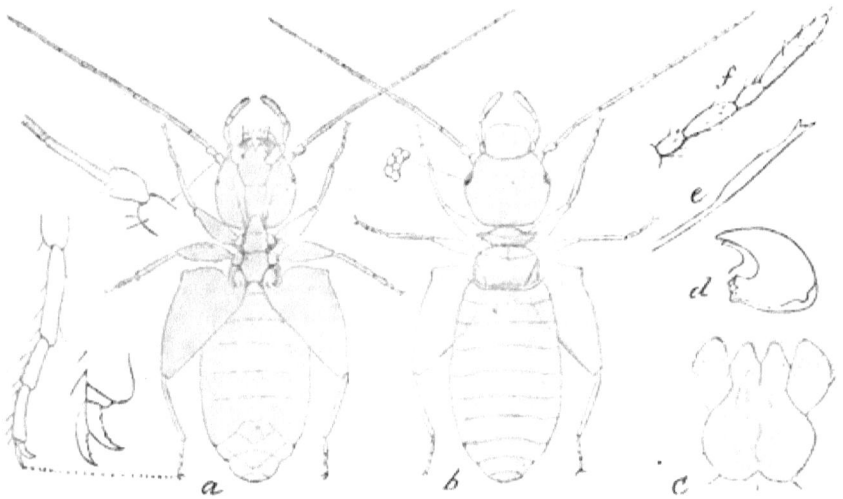

FIG. 34.—*Atropos divinatoria: a*, adult from below; *b*, same from above; *f*, maxillary palpus; *e*, maxilla (?); *d*, mandible; *c*, labium—all enlarged (original).

and may be taken as the type. All these troublesome house species are soft-bodied, wingless, degraded creatures, representing the very lowest form of insect life. A great many species, also, live out of doors, many of these being winged and somewhat resembling plant-lice. They frequently occur in numbers on the bark of trees and the walls of buildings, and feed on lichens or decaying vegetable matter. The Psocidæ are biting insects, having well-developed mandibles and other mouth parts.

One of the most interesting features in connection with the common house species, and from which it takes sometimes the name "death watch," is the reputation it has of making a ticking sound, supposed to prognosticate dire consequences to some inmate of the house. That it can make some such noise, probably by striking its head against some hard object, seems to be pretty well established in spite of the seeming

impossibility of an audible sound being produced in this way by so
small an insect. This psocid is not, however, the true deathwatch.
This doubtful honor is shared by a near ally, also a psocid, and having
similar habits (*Clothilla pulsatoria*), and certain wood-boring beetles,
which frequently work in the timbers of houses.

The house species, and particularly the one named at the head of
this chapter, are widely distributed, almost cosmopolitan, and are
occasionally the source of very considerable annoyance and damage.
Throughout the warm season they may be frequently seen in cupboards,
on window ledges, or library shelves, especially among books or papers
which are seldom used. They are practically omnivorous, feeding on
any animal or vegetable matter, and are especially fond of the starchy
paste used in book bindings or for attaching wall paper. They also
feed on flour, meal, and other farinaceous substances, and are frequently
very destructive to collections of natural history objects.

Under ordinary circumstances these insects are not especially injurious in dwelling houses, and it is only where the materials which they
are capable of injuring or in which they will breed are left undisturbed
for long periods that they are apt to multiply and cause any serious
damage. Occasionally, however, they will multiply in excessive numbers in some available food supply and swarm over the house, to the
great consternation of the housekeeper. In cases of such extraordinary multiplication, so difficult are they to reach in the many recesses
in which they can conceal themselves that the most persistent and
thorough cleansing and fumigating are scarcely of any avail. Fortunately, such instances of excessive multiplication are rare, but
there are several notable cases on record. The straw or husk fillings
of mattresses or beds seem to be especially favorable locations for their
multiplication, and in the worst cases of infestation the psocids have
come from such sources. Small species of psocids are often extraordinarily abundant in straw in barns and stables, and Dr. Lintner
quotes Mr. McLachlan, of London, England, as having found myriads
of the species under discussion in the straw coverings of wine bottles.

Mr. Alfred C. Stokes, Trenton, N. J. (Insect Life, Vol. I, p. 144),
reports a case which may be taken as a sample of several recorded
instances of a similar nature. He says that in a new house kept by
very neat occupants a mattress of hair and corn husks which had been
purchased some six months before was found in September, after the
house had been closed about six weeks, to be so covered with these
insects that "a pin point could not have been put down without touch-
ing one or more of the bugs." The side of the lower sheet next the
mattress was likewise covered, and further search showed the walls and
in fact the entire house to be swarming with them. A sweep of the
hand over the walls would gather them by thousands; bureau drawers
were swarming with them, and they were under every object and in
everything. The mattress was found to contain millions of them and

seemed to be the source of supply. The measures taken were most thorough. The mattress was promptly removed; walls and floors were washed with borax and corrosive sublimate solution; carpets were steam cleaned; pyrethrum was freely used; furniture was beaten, cleaned, and varnished, the struggle being kept up for a year with all the persistence of an extraordinarily neat housekeeper. The insect continued to have the best of it, however, and persisted, though in diminished numbers.

The family then removed to a hotel and for days the house was fumigated with burning sulphur and the scrubbing was repeated. The insect was still not entirely exterminated and the house was vacated again and subjected to the vapor of benzine. The insects, two years after the removal of the mattress, were reported to be still in the house, greatly reduced, but to be found in dark corners.

An almost exact duplication of this experience is reported by Dr. J. A. Lintner (Second Report, p. 198) as occurring in a residence in Otsego County, N. Y., the infestation coming originally from straw-filled ticks.

In aggravated cases of the kind noted nothing but the most thorough steps will be of avail. The source of supply, if in straw or husk ticks, should be promptly removed and the contents of the ticks or mattresses burned.

Carpets and bedding should be steam cleaned and floors should be thoroughly washed with soapsuds and the walls washed and repapered or painted. Benzine or gasoline should be applied freely to all possible retreats or to furniture which can not be otherwise cleaned. Thorough fumigation with brimstone, as recommended for the bedbug (see p. 38), or like fumigation with bisulphide of carbon, will destroy many of the psocids if the room can be tightly closed for several hours.

There is no means of preventing the occasional occurrence of psocids in houses, but unless exceptional opportunities are furnished they will rarely be troublesome, and occasional examinations of book shelves or other locations where they are apt to appear, with a liberal dusting of pyrethrum powder whenever necessary, will ordinarily keep them in check. With plenty of air and light and in apartments in daily use they rarely appear in any numbers. The use of straw or husk filled ticks or mattresses would seem inadvisable or at least should be discontinued at the first indication of being at all subject to infestation.

THE AMERICAN SPRING-TAIL.

(*Lepidocyrtus americanus* Marlatt.)

This very anomalous little insect, measuring scarcely more than one-tenth of an inch, silvery gray in color, with purple or violet markings, may be frequently observed in houses in situations similar to those frequented by the two species last described. In common with the silver fish. it belongs to the order of insects known as Aptera (wingless), from the fact of their having no vestige of wings throughout life.

The simple structure of these insects, and particularly their resemblance to the larval state of winged insects, has led to the belief that they are the primitive forms of insect life. That this is true is, however, by no means certain, and they may rather be degraded or debased examples of some of the higher orders of insects. The species figured herewith is not infrequently found in dwellings in Washington, but is apparently undescribed, and, in fact, little is known of the American species. It is, however, closely allied to a European form (*I. cerricalis*), often found in cellars, and figured by Sir John Lubbock in his monograph on these insects (Pl. XXV). Another allied European species (*Scira domestica*) has been named from the fact of its being a frequenter of houses.

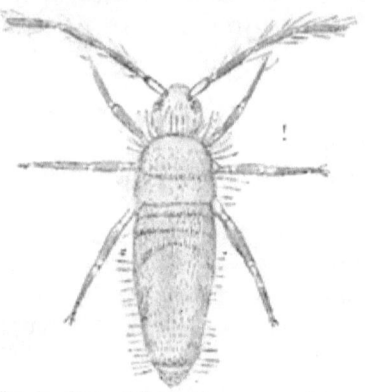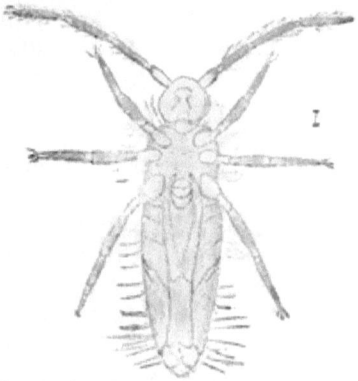

FIG. 35.—Spring-tail (*Lepidocystus americanus*) view from above (original).   FIG. 36.—Spring-tail (*Lepidocyrtus americanus*) view from beneath (original).

These insects belong to the suborder Collembola, which (following Sharp) is distinguished from the other suborder of Aptera, Thysanura, by having but five body segments instead of ten, and possessing a very peculiar ventral tube on the first segment, and commonly also a terminal spring, by means of which these creatures leap with great agility, and from which they take their common name of "spring-tails."

These insects, though very abundant, have been very little studied, and little is known of their life habits. They often multiply in extraordinary numbers, especially in moist situations, swarming on the surface of stagnant water or on wet soil. They seem to be very tolerant of cold, and we have interesting accounts of the occurrence of a species related to the one figured in the Arctic regions on melting snow fields and on glaciers, where they are known as "snow fleas" or "snow worms." Other interesting forms occur in caves, and in the Mammoth Cave in Kentucky they are notably abundant. In houses they may often be found on window sills, in bathrooms, and sometimes, under favorable situations, in very considerable numbers. Especially are they apt to occur where there are window plants or in small conservatories, but are not confined to these situations. Very little is known of their food habits, but they are supposed to subsist on refuse or chiefly decaying vegetable matter.

The striking peculiarities of these insects are in the remarkable ventral tube and the strong saltatorial appendage of the extremity of the body. The first arises from the forward body segment, and seems to act in this species as a sort of a retainer for the leaping organ, or spring proper, as shown in fig. 36. It is said to secrete a viscid fluid, which enables the insect to better adhere to smooth vertical

FIG. 37.—Spring-tail (*Lepidocyrtus americanus*). *a*, lateral view of female; *b*, foot of same; *c*, tip of spring-tail; *d*, body scale; *e*, upper lip or labrum; *f*, mandible or jaws; *g*, lower jaws and lower lip or maxillæ and labium—(original).

surfaces. The so-called "catch," or retainer proper, is shown in a small projection between the hind pair of legs and the spring (fig. 37), and grasps the latter near the middle. The springing organ is two-jointed, the last joint being bifurcate, and its terminals inclosing the ventral tube. It is shown in normal position in fig. 36, and as it appears when leaping in fig. 37, *a*.

These insects can not survive dryness, and, while they will not often occur in sufficient numbers to be particularly objectionable, the removal of the moist objects or surfaces on which they congregate and the maintenance of dry conditions will cause them to soon disappear.

CHAPTER VI.

COCKROACHES AND HOUSE ANTS.

By C. L. MARLATT.

COCKROACHES.

(Periplaneta americana et al.)

Roaches are among the commonest and most offensive of the insects which frequent human habitations. They were well known to the ancients, who called them *lucifuga*, from their habit of always shunning the light. The common English name for them, or, more properly, for the common domestic English species, is "black beetle." In America this name has not been adopted to any extent for this insect, which was

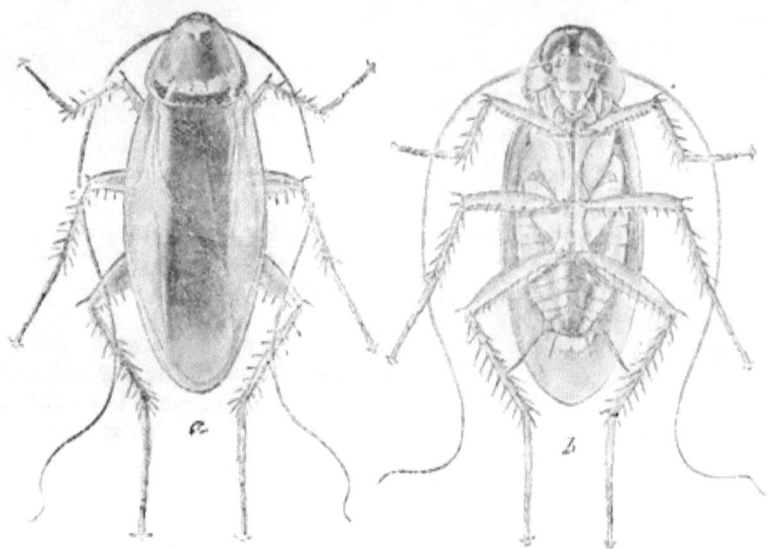

FIG. 38.—The American roach (*Periplaneta americana*): *a*, view from above; *b*, from beneath—both enlarged one-third (original).

early introduced here, and the term "roach," or "cockroach," is the common appellation of all the domestic species. The little German roach, however, is very generally known as the Croton bug, from its early association with the Croton waterworks system in New York City. The popular designations of this insect in Germany illustrate in an

84

amusing way both sectional and racial prejudices. In north Germany
these roaches are known as "Schwaben," a name which applies to the
inhabitants of south Germany, and the latter section "even up" by call-
ing them "Preussen," after the north Germans. In east Germany they
are called "Russen," and in west Germany "Franzosen," the two latter
appellations indicating a certain national antipathy to rival countries
as well as a fanciful idea as to origin. Still other names are "Spanier,"
dating from the time of Charles V, and "Däne," from Denmark.

DISTRIBUTION AND HISTORY.

The roaches belong to a very extensive family, the Blattidæ, comparatively few of which, fortunately, have become domesticated. In
temperate countries some four or five species are very common household pests, and a few occur wild in woods; but they are essentially
inhabitants of warm countries, and in the Tropics the house species are
very numerous, and the wild species occur in great number and variety,
many of them being striking in shape, coloration, and size, one species
expanding more than 6 inches. The inability of the domestic roaches
to withstand unusual cold was illustrated by the fact that the severe
weather in the winter of 1894 in Florida, which was so destructive to
the citrus groves, on the authority of Mr. H. G. Hubbard, destroyed
all the roaches, even those in houses, except a few unusually well protected. Under suitable conditions in the more northern latitude the
domestic species often multiply prodigiously, and even in the far north
a species occurs in the huts of the Laplanders, and sometimes entirely
devours the stores of dried fish put away for winter consumption.

While the domestic species are few in number, nearly a thousand
species of Blattidæ have been described and preserved in collections,
and it is estimated that perhaps upward of 5,000 species occur at the
present time in different parts of the world. The great majority of the
roaches live out of doors, subsisting on living vegetation, and occasionally in warm countries are very injurious to cultivated plants.

The roach is one of the most primitive and ancient insects, in the
sense of its early appearance on the globe, fossil remains of roaches
occurring in abundance in the early coal formations, ages before the
more common forms of insect life of the present day had begun to
appear. The species now existing are few in number in comparison
with the abundance of forms in the Carboniferous age, which might
with propriety be called the age of cockroaches, the moisture and
warmth of that distant period being alike favorable to plant growth
and the multiplication of this family of insects.

The house roaches of today were undoubtedly very early associated
with man in his primitive dwellings, and through the agency of commerce have followed him wherever navigation has extended. In fact,
on shipboard they are always especially numerous and troublesome,
the moisture and heat of the vessels being particularly favorable to

their development. It is supposed that the common oriental cockroach or so-called "black beetle" of Europe (*Periplaneta orientalis*) is of Asiatic origin, and it is thought to have been introduced into Europe in the last two or three hundred years. The original home of this and the other common European species (*Ectobia germanica*) is, however, obscure, and in point of fact they have probably both been associated with man from the earliest times, and naturally would come into the newly settled portions of Europe from the older civilizations of Asia and Egypt.

Of the other two domestic species especially considered in this paper the Australian roach (*P. australasiæ*), as its name implies, is a native of Australia, and the American roach (*P. americana*) of tropical America.

Rarely do two of the domestic species occur in any numbers together in the same house. Often also of two neighboring districts one may be infested with one species, while in the other a distinct species is the commoner one. The different species are thus seemingly somewhat antagonistic, and it is even supposed that they may prey upon each other, the less numerous species being often driven out.

STRUCTURAL CHARACTERISTICS.

Although among the oldest insects geologically, roaches have not departed notably from the early types, and form one of the most persistent groups among insects. The house species are rather uniformly dark brown or dark colored, a coloration which corresponds with their habit of concealment during daylight. They are smooth and slippery insects, and in shape broad and flattened. The head is inflexed under the body, so that the mouth parts are directed backward and the eyes directed downward, conforming with their groveling habits. The antennæ are very long and slender, often having upward of 100 joints. The males usually have two pairs of wings, the outer ones somewhat coriaceous and the inner ones more membranous and once folded longitudinally. In some species, as, for instance, the black beetle, the females are nearly wingless. The legs are long and powerful and armed with numerous strong bristles or spines. The mouth parts are well developed and with strong biting jaws, enabling them to eat all sorts of substances.

HABITS AND LIFE HISTORY.

In houses roaches are particularly abundant in pantries and kitchens, especially in the neighborhood of fireplaces, on account of the heat. For the same reason they are often abundant in the oven rooms of bakeries or wherever the temperature is maintained above the normal. They conceal themselves during the day behind baseboards, furniture, or wherever security and partial protection from the light are afforded. Their very flat, thin bodies enable them to squeeze themselves into small cracks or spaces where their presence would not be suspected and where they are out of the reach of enemies. Unless routed out by the moving of furniture or disturbed in their hiding places, they are rarely seen, and if so uncovered, make off with wonderful celerity, with a scurrying, nervous gait, and usually are able to elude all efforts at their capture or destruction. It may often happen that their presence, at least in the abundance in which they occur, is hardly realized by the housekeeper, unless they are surprised in their midnight feasts. Coming into a kitchen or pantry suddenly, a sound of the rustling of numerous objects will come to the ear, and if a light be introduced, often the floor or shelves will be seen covered with scurrying roaches hastening to places of concealment. In districts where the large American roach occurs they sometimes swarm in this way at night in such numbers that upon entering a small room in which they are congregated one will be repeatedly struck and scratched on the face and hands by the insects in their frantic flight to gain concealment.

The black roach is less active and wary than the others, and particularly the German roach, which is especially agile and shy.

The domestic roaches are practically omnivorous, feeding on almost any dead animal matter, cereal products, and food materials of all sorts. They are also said to eat their own cast skins and egg cases, and it is supposed that they will attack other species of roaches, or are, perhaps, occasionally cannibalistic. They will also eat or gnaw woolens, leather (as of shoes or furniture), and frequently are the cause of extensive damage to the cloth and leather bindings of books in libraries and publishing houses. The sizing or paste used on the cloth covers and in the binding of books seems to be very attractive. The surface of the covers of cloth-bound books is often much scraped and disfigured, particularly by the German cockroach (*Ectobia germanica*), and the gold lettering is sometimes eaten off to get at the albumen paste. On shipboard the damage is often very extensive, on account of the vast numbers of cockroaches which frequently occur there, and we have reliable accounts of entire supplies of ship biscuits having been eaten up or ruined by roaches.

The damage they do is not only in the products actually consumed, but in the soiling and rendering nauseous of everything with which they come in contact. They leave, wherever they occur in any numbers, a fetid, nauseous odor, well known as the "roachy" odor, which is persistent and can not be removed from shelves and dishes without washing with soap and boiling water. Food supplies so tainted are beyond redemption. This odor comes partly from their excrement, but chiefly from a dark-colored fluid exuded from the mouth of the insect, with which it stains its runways, and also in part, doubtless, from the scent glands, which occur on the bodies of both sexes between certain segments of the abdomen, and which secrete an oily liquid possessing a very characteristic and disagreeable odor. It frequently happens that shelves on which dishes are placed become impregnated with this roachy odor, and this is imparted to and retained by dishes to such an extent that everything served in them, particularly liquids, as coffee or tea, will be noticed to have a peculiar, disgusting, foreign taste and odor, the source of which may be a puzzle and will naturally be supposed to come from the food rather than from the dish.

The roaches are normally scavengers in habit and may at times be of actual service in this direction by eating up and removing any dead animal material.

One other redeeming trait has been recorded of them, namely, that they will prey upon that other grievous pest of houses which are not subjected to careful supervision, the bedbug. Their habits in this direction have been recorded several times. One writer, in a narrative of a voyage (Foster's Voyage, Vol. I, p. 373), makes the following statement in this connection:

Cockroaches, those nuisances to ships, are plentiful at St. Helena, and yet, bad as they are, they are more endurable than bugs. Previous to our arrival here in the *Chanticleer*, we had suffered great inconvenience from the latter, but the cockroaches no sooner made their appearance than the bugs entirely disappeared. The fact is that the cockroach preys upon them and leaves no sign or vestige of where they have been. So that it is a most valuable insect.[1]

The cockroach is, however, far too much of a nuisance itself to warrant its being recommended as a means of eradicating even the much more disagreeable insect referred to.[2]

The local spread of roaches from house to house is undoubtedly often effected by their being introduced with supplies, furniture, goods, etc. That the Croton bug, or German roach, and probably the other species also, may develop a migratory instinct has been witnessed by Dr. Howard and the writer in Washington. (See Insect Life, Vol. VII, p. 349.)

This very interesting instance of what seems to have been a true migration, in which an army of thousands of roaches by one common impulse abandoned their old quarters and started on a search for a more favorable location, illustrates, as pointed out by Dr. Howard, what is probably of frequent occurrence under the cover of darkness, and accounts for the way in which new houses frequently become suddenly overrun with these vermin.

---

[1] Proc. Ent. Soc. Lond., 1855, N. S. 3, p. 77.

[2] The following interesting letter from Mr. Herbert H. Smith, the collector and naturalist, gives a vivid picture of the roach nuisance in the Tropics:

"Cockroaches are so common in Brazilian country houses that nobody pays any attention to them. They have an unpleasant way of getting into provision boxes, and they deface books, shoes, and sometimes clothing. Where wall paper is used they soon eat it off in unsightly patches, no doubt seeking the paste beneath. But at Corumba, on the upper Paraguay, I came across the cockroach in a new rôle. In the house where we were staying there were nearly a dozen children, and every one of them had their eyelashes more or less eaten off by cockroaches—a large brown species, one of the commonest kind throughout Brazil. The eyelashes were bitten off irregularly, in some places quite close to the lid. Like most Brazilians, these children had very long, black eyelashes, and their appearance thus defaced was odd enough. The trouble was confined to children, I suppose because they are heavy sleepers and do not disturb the insects at work. My wife and I sometimes brushed cockroaches from our faces at night, but thought nothing more of the matter. The roaches also bite off bits of the toe nails. Brazilians very properly encourage the large house spiders, because they tend to rid the house of other insect pests."

LIFE HISTORY.

The roach in its different stages from egg to adult shows comparatively little variation in appearance or habits. The young are very much like the adult, except in point of size and in lacking wings, if the latter be winged in the adult state. In their mode of oviposition they present, however, a very anomalous and peculiar habit. The eggs, instead of being deposited separately, as with most other insects, are brought together within the abdomen of the mother into a hard, horny pod or capsule which often nearly fills the body of the parent. This capsule contains a considerable number of eggs, the number varying in the different species, arranged in two rows, the position of the eggs being indicated on the exterior of the capsule by transverse lateral impressions. When fully formed and charged with eggs the capsule is often partly extruded from the female abdomen and retained in this position sometimes for weeks, or until the young larvæ are ready to emerge. The capsule is oval, elongate, or somewhat bean shaped, and one of its edges is usually serrate. The young are in some instances assisted to

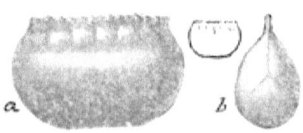

Fig. 39.—Egg-capsule of *Periplaneta americana: a*, side; *b*, end view.—natural size indicated by outline figure (original).

escape by the parent, who with her feet aids in splitting the capsule on the serrate edge to facilitate their exit. On hatching, it is said, the young are often kept together by the parent and brooded over and cared for, and at least a colony of young will usually be found associated with one or two older individuals. These insects are more or less gregarious, notably so in the case of the black beetle of Europe and to a less extent with the German and American roaches.

They pass through a variable number of molts, sometimes as many as seven, the skin splitting along the back and the insects coming out white, soft, but rapidly hardening and assuming the normal color. Some astounding statements have been made as to the length of time required for the development of the roach from the egg to the adult. Four or five years have been said to be necessary for an individual to reach full growth; but more recent breeding experiments have not altogether confirmed these statements. Their development, however, is unquestionably slow, and probably under the most favorable conditions rarely is more than one generation per year produced. In colder countries the breeding and growth are practically restricted to the warm season. During the winter months they go into concealment and partial hibernation. *Ectobia germanica* has been shown to reach full growth in a variable period from four and a half to six months (Hummel, Essais Entomologiques, No. 1, St. Petersburg, 1821). The common American roach (*Periplaneta americana*) has been carried from the egg to the adult state in our insectary. Young hatching July 11

from an egg case received from Eagle Pass, Tex., reached the adult
stage between March 14 and June 12 of the following year, indicating
a period of nearly twelve months for complete development. The rate
of growth of the roach undoubtedly depends very largely on food and
temperature, and under unfavorable conditions the time required for
development may undoubtedly be vastly lengthened. The abundance
of roaches is, therefore, apparently not accounted for so much by their
rapidity of multiplication as by their unusual ability to preserve
themselves from ordinary means of destruction and by the scarcity of
natural enemies.

THE COMMON DOMESTIC ROACHES.

The four roaches which have been made the subject of illustrations
represent the species which occur most commonly in houses, bakeries,
or on shipboard. The numerous tropical house species, many of which
are perhaps only partially domesticated, and the subarctic roach of high
altitudes and of the extreme north have been omitted.

The American roach (*Periplaneta americana*) (fig. 38) is the native or
indigenous species of this continent, originating, it is supposed, in tropical or subtropical America.

The ancient and rather quaint account of this insect[1] quoted below
in a footnote indicates that this species early came to the notice
of our forefathers. Its domesticity doubtless resulted from ages of
association with the aborigines. It has now become thoroughly cosmopolitan, and is unquestionably the most injurious and annoying of the
species occurring on vessels. It is sometimes numerous also in greenhouses, causing considerable injury to tender plants. It is a notorious
house pest and occasionally vies with the German roach in its injuries
to book bindings. One of the most serious cases of injury of this sort
was reported by the Treasury Department. The backs, sometimes
entirely, of both cloth and leather bound books were eaten off to get
at the starchy paste used in the binding. (Insect Life, Vol. I, p. 67-70.)

It is very abundant in the Middle and Western States, where it has
been until recently practically the only troublesome house species. In
the East it is not often so common as are one or other of the following
species and especially *germanica*. In foreign countries it has not become
widespread and is largely confined to seaport towns. In size it is

---

[1] *The cockroach.*—These are very troublesome and destructive vermin, and are so
numerous and voracious, that it is impossible to keep victuals of any kind from
being devoured by them, without close covering. They are flat, and so thin that
few chests or boxes can exclude them. They eat not only leather, parchment and
woollen, but linen and paper. They disappear in Winter, and appear most numerous in the hottest days in Summer. It is at night they commit their depredations,
and bite people in their beds, especially children's fingers that are greasy. They lay
innumerable eggs, creeping into the holes of old walls and rubbish, where they
lie torpid all the Winter. Some have wings, and others are without—perhaps of
different Sexes. (Catesby: Nat. Hist. Carolina, 1748, Vol. II, p. 10.)

larger than any of the other domestic species, and it is light brown in color, the wings being unusually long, powerful, and well developed in both sexes.

The Australian roach (*Periplaneta australasiæ*) resembles very closely the last species, but differs strikingly in the brighter and more definitely limited yellow band on the prothorax and in the yellow dash on the sides of the upper wings (see fig. 40). In the United States it is the most abundant and troublesome species in Florida and some of the other Southern States. It is already practically cosmopolitan.

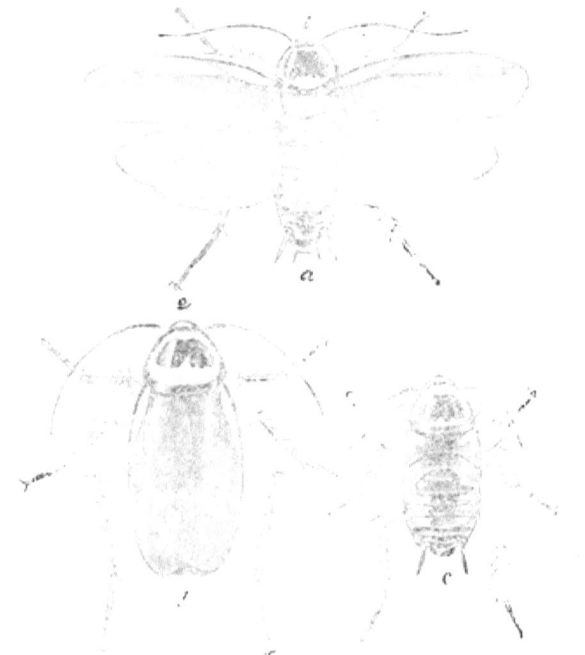

FIG. 40.—The Australian roach (*Periplaneta australasiæ*): *a*, male with spread wings; *b*, female; *c*, pupa—all life size (original).

The oriental cockroach, or black beetle (*Periplaneta orientalis*), is the common European and particularly the English species, and is notable for the fact that the female is nearly wingless in the adult state. The wings of the male also are shortened, not reaching to the extremity of the body. In color it is very dark brown, almost black, shining, and rather robust, much stouter than the other species, making its English name of "black beetle" quite appropriate. This species is notably gregarious in habit, individuals living together in colonies in the most amicable way, the small ones being allowed by the larger ones to sit on them, run over them, and nestle beneath them without any resentment being shown. This species was a common and troublesome pest in the British colonies early in the eighteenth century, although unknown at the same time in the French Canadian possessions.[1]

It then seemed to be commonly known as the mill beetle. The early Dutch called them *Kakerlach*, and in the Swede settlements they were known as *Brodætare*

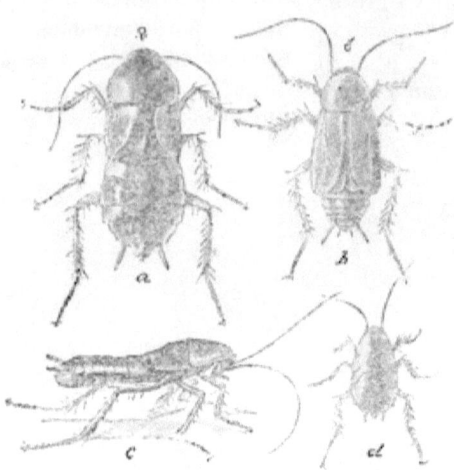

(bread eaters). It is now very common in houses in the East, but is quite generally distributed, and is the common species even so far removed from the Atlantic seaboard as New Mexico. The characteristics of this insect are shown in the accompanying illustration (fig. 41).

The German cockroach, *Ectobia* (*Phyllodromia*) *germanica*, is particularly abundant in Germany and neighboring European countries, but, like most of the other domestic species,

FIG. 41.—The oriental roach (*Periplaneta orientalis*): *a*, female; *b*, male; *c*, side view of female; *d*, half-grown specimen—all natural size (original).

has become world-wide in distribution. In this country it is very often styled the Croton bug, this designation coming from the fact, already alluded to, that attention was first prominently drawn to it at the time of the completion of the Croton system of waterworks in New York City. It had probably been introduced long previously, but the extension of the waterworks system and of piping afforded it means of ingress into residences, and greatly encouraged its spread and facilitated its multi-

FIG. 42.—The German roach (*Ectobia germanica*): *a*, first stage; *b*, second stage; *c*, third stage; *d*, fourth stage; *e*, adult; *f*, adult female with egg-case; *g*, egg-case—enlarged; *h*, adult with wings spread—all natural size except *g*. (From Riley.)

plication. The dampness of water pipes is favorable to it, and it may be carried by the pressure of the water long distances through the pipes without injury. This roach has so multiplied in the eastern United States that it has now become the commonest and best known of the domestic species, and its injuries to food products, books, etc., and the disgusting results of its presence in pantries, storehouses, and bakeries, give it really a greater economic importance than any of the other species.

It is very light brown in color, and characteristically marked on the thorax with two dark-brown stripes. It is more active and wary than the larger species and much more difficult to eradicate. It is the smallest of the domestic species, rarely exceeding five-eighths of an inch in length, and multiplies much more rapidly than the others, the breeding period being shorter and the number of eggs in the capsules greater than with the larger roaches. The injuries effected by it to cloth-bound reports have been the source of very considerable annoyance at the Department of Agriculture and in the large libraries of Eastern towns and colleges. The characteristics of the different stages, from the egg to the adult, are shown in the illustration (fig. 42).

REMEDIES.

Like the crows among birds, the roaches among insects are apparently unusually well endowed with the ability to guard themselves against enemies, displaying great intelligence in keeping out of the way of the irate housekeeper and in avoiding food or other substances which have been doctored with poisons for their benefit. Their keenness in this direction is unquestionably the inheritance of many centuries during which the hand of man has ever been raised against them.

The means against these insects, including always vigilance and cleanliness as important preventives, are three, namely, destruction by poisons, by fumigation with poisonous gases, and by trapping.

Poisons.—As just noted, roaches often seem to display a knowledge of the presence of poisons in food, and, notwithstanding their practically omnivorous habits, a very little arsenic in baits seems to be readily detected by them. In attempting to eradicate roaches from the Department storerooms where cloth-bound books are kept various paste mixtures containing arsenic were tried, but the roaches invariably refused to feed on them in the least. This applies particularly to the German roach, or Croton bug, and may not hold so strongly with the less wary and perhaps less intelligent larger roaches.

A common remedy suggested for roaches consists in the liberal use of pyrethrum powder or buhach, and when this is persisted in considerable relief will be gained. It is not a perfect remedy, however, and is at best but a temporary expedient, while it has the additional disadvantage of soiling the shelves or other objects over which it is dusted. When used it should be fresh and liberally applied. Roaches are often paralyzed by it when not killed outright, and the morning after an application the infested premises should be gone over and all the dead or partially paralyzed roaches swept up and burned.

There are many proprietary substances which claim to be fairly effective roach poisons. The usefulness of most of these is, however, very problematical, and disappointment will ordinarily follow their application. The only one of these that has given very satisfactory results is a phosphorous paste, also sold in the form of pills. It probably consists of sweetened flour paste containing phosphorus, and is spread on bits of paper or cardboard and placed in the runways of the roaches. It has been used very successfully in the Department to free desks from Croton bugs, numbers of the dead insects being found in the drawers every day during the time the poison was kept about.

Fumigation.—Wherever roaches infest small rooms or apartments which may be sealed up nearly air-tight, and also on shipboard, the roach nuisance can be greatly abated by the proper use of poisonous gases, notably bisulphide of carbon. This substance, distributed about a pantry or room in open vessels, will evaporate, and, if used in sufficient quantity, will destroy roaches. Unless the room can be very tightly sealed up, however, the vapor dissipates so rapidly that its effect will be lost before the roaches are killed. The hatches of ships, especially of smaller coasting vessels, may be battened down, a very liberal application of bisulphide of carbon having been previously made throughout the interior. If left for twenty-four hours the roaches and all other vermin will unquestionably have been destroyed. In the use of this substance it must be always borne in mind that it is violently explosive in the presence of fire, and every possible precaution should be taken to see that no fire is in or about the premises during the treatment. It is also deadly to higher animals, and compartments should be thoroughly aired after fumigation. A safer remedy of the same nature consists in burning pyrethrum in the infested apartment. The smoke and vapors generated by the burning of this insecticide are often more effective in destroying roaches than the application of the substance in the ordinary way as a powder. There is no attendant danger of explosion, and the only precaution necessary is to see that the room is kept tightly closed for from six to twelve hours. The smoke of burning gunpowder is also very obnoxious and deadly to roaches, particularly the black English roach. On the authority of Mr. Theo. Pergande, gunpowder is commonly used in Germany to drive these roaches out of their haunts about fireplaces. The method consists in molding cones of the moistened powder and placing them in the empty fireplace and lighting them. The smoke coming from the burning powder causes the roaches to come out of the crevices about the chimney and fire bricks in great numbers, and rapidly paralyzes or kills them, so that they may be afterwards swept up and destroyed. This remedy will only apply to old houses with large fireplaces, and has no especial significance for the modern house. It is presented, however, as a means applicable wherever conditions similar to those described occur.

Trapping.—Various forms of traps have been very successfully employed in England and on the continent of Europe as a means of collecting and destroying roaches. These devices are all so constructed that the roaches may easily get into them and can not afterwards escape. The destruction of the roaches is effected either by the liquid into which they fall or by dousing them with hot water. A few of the common forms of traps and the methods of using them are here described.

A French trap consists of a box containing an attractive bait, the cover of which is replaced by four glass plates inclined toward the center. The roaches fall from the covering glasses into the box and are unable to escape. A similar trap used in England is described by Westwood. It consists of a small wooden box in which a circular hole is cut in the top and fitted with a glass ring, so that it is impossible for the roaches to escape. This trap is baited nightly, and the catch thrown each morning into boiling water. A simpler form of trap, which I am informed by Mr. F. C. Pratt is very successfully used in London, England, consists of any deep vessel or jar, against which a number of sticks are placed, and bent over so that they project into the interior of the vessel for a few inches. The vessel is partially filled with stale beer or ale, a liquid for which roaches seem to have a special fondness. In the morning these vessels are found charged with great quantities of dead and dying roaches, which have climbed up the inclined sticks and slipped off into the vessel. We have had fair success with this last method against the oriental roach in Washington, but against the more wary and active Croton bug it seems less effective.

Traps of the sort described, placed in pantries or bakeries, will unquestionably destroy great quantities of roaches, and keep them, perhaps, more effectively in check than the use of the troublesome insect powders or the distribution of poisoned bait, especially as the latter are so often ineffective.

NATURAL ENEMIES AND PARASITES.

The common European egg parasite of the roach, *Evania appendigaster*, is now probably widely distributed. It occurs in the United States and has also been found in Cuba. Unfortunately, its usefulness is largely impaired by the occurrence of a secondary parasite, *Entedon hagenowi*, which preys upon and destroys the first, and has also been introduced into this country with it.

A correspondent informs us also that the common tree frog will clear rooms of roaches over night very effectually.

HOUSE ANTS.

(*Monomorium pharaonis, et al.*)

There are a number of species of ants often occurring in houses, the more important of which are common to both hemispheres, and are probably of Old World origin. One of these, the little red ant (*Monomorium pharaonis* Linn.), has become thoroughly domesticated and passes its entire existence in houses, having its nests in the walls or beneath the flooring, and usually forming its new colonies in similar favorable situations. Two other ants are very common nuisances in houses, namely, the little black ant (*Monomorium minutum* Mayr) and the pavement ant of the Atlantic Seaboard (*Tetramorium cæspitum* Linn.). None of these ants are so destructive to household effects or supplies as they are annoying from the mere fact of their presence and their faculty for "getting into" articles of food, particularly sugars, sirups, and other sweets. Having once gained access to stores of this sort, the news of the discovery is at once conveyed to the colony, and in an incredibly short time the premises are swarming with these unwelcome visitors.

In habits and life history these ants are all much alike, and, in common with other social insects, present that most complex and interesting phase of communal life, with its accompanying division of labor and diversity of forms of individuals, all working together in the most

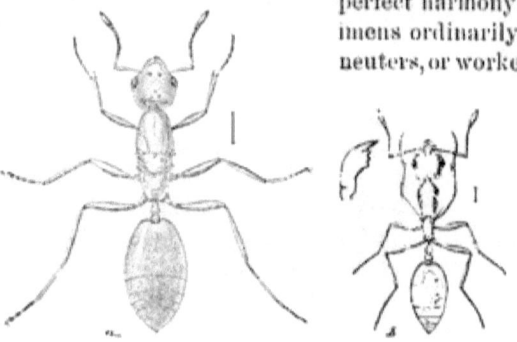

perfect harmony and accord. The specimens ordinarily seen in houses are all neuters, or workers. In the colony itself, if it be discovered and opened, will be found also the larger wingless females and, at the proper season, the winged males and females. During most of the year, however, the colony consists almost exclusively of workers, with one or

Fig. 43.—The red ant (*Monomorium pharaonis*): *a*, female; *b*, worker—enlarged (from Riley).

more perfect wingless females. Winged males and females are produced during the summer and almost immediately take their nuptial flight. The males soon perish, and the females shortly afterwards tear off their own wings, which are but feebly attached, and set about the establishment of new colonies. The eggs, which are produced in extraordinary numbers by the usually solitary queen mother, are very minute, oval, whitish objects, and are cared for by the workers, the young larvæ being fed in very much the same way as in the colonies of the hive bee. The so-called ant eggs, in the popular conception, are not eggs at all, but the white larvæ and pupæ, and, if of females or males, are much larger than the workers and many times larger than the true eggs.

As a house species the red ant (*Monomorium pharaonis* Linn.) (fig. 43) is the common one. It is practically cosmopolitan, and its exact origin is unknown. This species, nesting habitually in the walls of houses or beneath flooring, is often difficult to eradicate. There is no means of doing this except to locate the nest by following the workers back to their point of entrance. If in a wall the inmates may sometimes be reached by injecting bisulphide of carbon or a little kerosene. If under flooring it may sometimes be possible to get at them by taking up a section. Unless the colony can be reached and destroyed all other measures will be of only temporary avail.

The little black ant (*Monomorium minutum* Mayr) (fig. 44) is not strictly a house species. although frequently occurring indoors, and becoming at times quite as troublesome as the red ant. Its colonies usually occur under stones in yards, but are frequently found in the fields, and will be recognized from the little pyramids of fine grains of soil which sur-

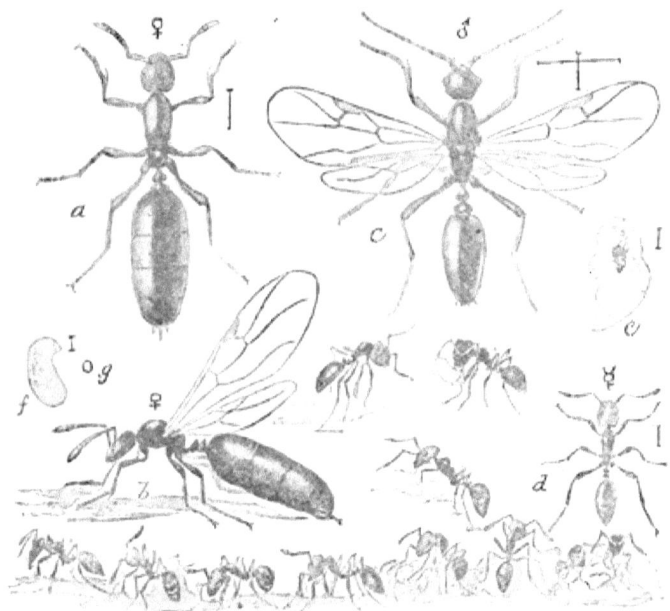

FIG. 44.—The little black ant (*Monomorium minutum*): *a*, female; *b*, same with wings; *c*, male; *d*, workers; *e*, pupa; *f*, larva; *g*, egg of worker—all enlarged (original).

round the entrances to the excavations. If these colonies be opened they will be found to contain workers and usually one or more very much larger gravid females. This species, when occurring in houses, can often be traced to its outdoor colony, and the destruction of this will prevent further trouble.

The pavement ant of our Eastern cities (*Tetramorium cæspitum* Linn.) (fig. 45) is in Europe the common meadow ant, and is two or three times larger than either of the other species referred to. It was early introduced into this country, and, while not yet reported from the West, is very common in Eastern towns, and particularly here in Washington. It has readily accommodated itself to the conditions of urban existence,

2805—No. 4——7

and commonly has its colonies under pavements, where it is often difficult of access. or beneath flagging or stones in yards.   It is often a more
persistent and pestilent house nuisance than the true house ant.

This seems to be the species referred to by Kalm[1] in 1748 as often
occurring in houses in Philadelphia and manifesting a great fondness
for sweets.   He records also some interesting experiments made by
Mr. (Benjamin?) Franklin, indicating the ability of these ants to commu-
nicate with one another.

The colonies of the pavement ant are often large, and they may frequently be uncovered in masses of a quart or more on turning over
stones in yards or lifting flagging in paths.

This ant may be often with little difficulty traced to its nest, which,
if accessible, or not thoroughly protected by unbroken pavement, as of

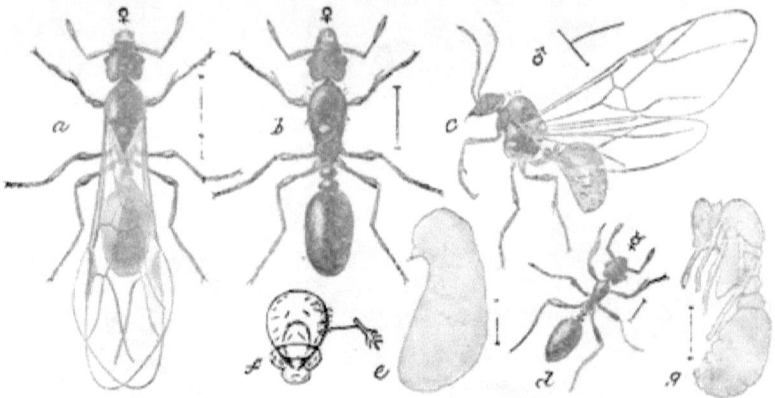

Fig. 45.—The pavement ant (*Tetramorium cœspitum*): *a.* winged female; *b.* same without wings;
c. male; *d.* worker; *e.* larva of female; *f.* head of same; *g.* pupa of same—all enlarged (original).

asphalt, can be rather easily exterminated.   So well established is the
species, however, that new colonies will usually soon take the place of
those destroyed.

Drenching the nests with boiling water or saturating them with coal
oil, which latter also may be introduced into cracks in pavements or
walls, are effective means of abating the nuisance of this ant.

There are several other ants closely resembling this last, mostly
species of *Lasius*, some foreign and some native, which form large
colonies in yards, throwing up earthen ant hills, beneath which are
extensive systems of underground galleries.   These may often get into
near-by houses and become quite as troublesome as the ants already
mentioned.

Excellent success has been had in destroying these ants with the use
of bisulphide of carbon applied in their nests.   The method consists in
pouring an ounce or two of the bisulphide into each of a number of

[1] Kahn's Travels, Vol. I, p. 238.

holes made in the nest with a stick, promptly closing the holes with the foot. The bisulphide penetrates through the underground tunnels and kills the ants in enormous numbers, and if applied with sufficient liberality will exterminate the whole colony.

Whenever the nests of any of these ants can not be located, there is no other resource but the temporary expedient of destroying them wherever they occur in the house. The best means of effecting this end is to attract them to small bits of sponge moistened with sweetened water and placed in the situations where they are most numerous. These sponges may be collected several times daily and the ants swarming in them destroyed by immersion in hot water. It is reported also that a sirup made by dissolving borax and sugar in boiling water will effect the destruction of the ants readily and in numbers. The removal of the attracting substances, wherever practicable, should always be the first step.

CHAPTER VII

SOME INSECTS AFFECTING CHEESE, HAMS, FRUIT, AND VINEGAR.

By L. O. HOWARD.

THE CHEESE, HAM, AND FLOUR MITES.

(*Tyroglyphus longior* L. and *T. siro* Gerv.)

Very minute, more or less colorless, eight-legged creatures swarm in numbers over and in old cheese and various stored products, such as dried meats, dried fruit, vanilla, and flour of different kinds. The species may be distinguished by the illustrations. *Tyroglyphus longior*

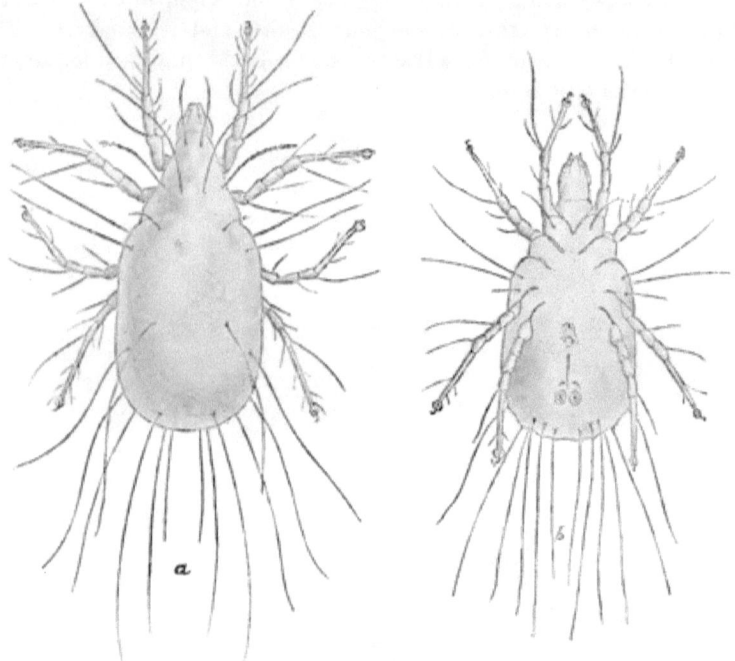

FIG. 46.—*Tyroglyphus longior:* a, female; b, male—greatly enlarged (after Canestrini).

is more rapid in its movements, larger in size, with longer and more cylindrical body, and longer and more numerous shining hairs sticking out on the sides. The two species are frequently found feeding in common.

Both species are common to Europe and the United States, and both have probably been carried to all parts of the world in food supplies.

Aristotle knew the cheese mites and spoke of them as the smallest of living creatures. Many subsequent writers have figured them and mentioned them, but the full life history was not known until 1868, when Claparède determined that the genus Hypopus was composed of forms which are steps in the development of true tyroglyphids.

All through the summer months, and in warm houses during the winter months, these creatures breed with astonishing rapidity and fecundity. The rapidity of multiplication and the extraordinary numbers in which these mites will occur under favorable conditions are almost incredible. In 1882 *T. longior* was found in an Ohio packing house, covering the dried and packed refuse (ready for sale as a fertilizer) in a layer which in some places was half an inch in thickness. At a low estimate 1 square inch of such a layer would contain 100,000 individuals. The females bring forth their young alive, and these in turn reach full growth and reproduce, until a cheese, once infested by a few, swarms with the crawling multitude, which cause its solid mass to crumble and become mixed with excremental pellets and cast-off skins. Through the summer months the mites are soft bodied and have comparatively feeble powers of locomotion, and when they have become numerous enough to devour the whole of a cheese, with no other food at hand, it was for a long time a puzzle to know what became of them and to understand how a cheese could become affected without contact with another infested cheese or without being placed in an infested room. It has been ascertained, however, that when necessity requires it, and when the insects happen to be in the proper stage of growth,

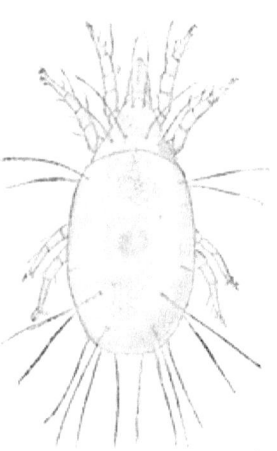

FIG. 47.—*Tyroglyphus siro.* female—greatly enlarged (after Berlese).

they have the power of not only almost indefinitely prolonging existence, but of undergoing a complete change of form, acquiring hard, brown protective coverings into which all of the legs can be drawn in repose. Back in Van Leeuwenhoek's time this Dutch naturalist showed that even the softer form can undergo a fast of eleven weeks without apparent discomfort, and it is now known that in the hard-shell or Hypopus state it may remain for many months without food.

In the majority of cases, however, where a given cheese is completely destroyed, all of the young and old mites perish and only those of middle age which are ready to take on the Hypopus condition survive. These fortunate survivors, possessing their souls with patience, retire into their shells and fast and wait, and as everything comes to him who waits, some lucky day a mouse or house fly or some other insect comes that way, and the little mite clings to it and is carried away to some spot where another cheese or food in some other form is at hand. It is in this way, as well as by the more readily understood means, that new cheese becomes infested and that the insect makes its appearance in pantries supposed to be perfectly clean.

REMEDIES AND PREVENTIVE MEASURES.

When we consider the great hardihood and extreme tenacity of life of this insect in the Hypopus condition, and the fact that almost every flying or crawling thing may become its common carrier, the difficulty of disinfecting a storeroom and of keeping it disinfected becomes very plain. Nothing, in fact, but the utmost cleanliness and watchfulness will prevent the appearance of the mites. When they have once entered a cheese, for example, there is no remedy except to cut out the infested portions. All energies must be bent toward prevention. If a given room seems to be badly infested it should be cleaned out, fumigated with sulphur, and washed out thoroughly with kerosene emulsion. Food supplies liable to be infested should be inspected daily during hot weather.

It is a point of considerable interest and of some practical account that there often occur, where these mites are present in numbers, one or more species of predaceous mites which feed exclusively on the injurious individuals and tend to greatly lessen their numbers. Some years ago a gentleman in Milwaukee sent the writer some thousands of mites which were found in a bin of wheat in an old elevator. They occurred in such numbers that every morning a quart or more could be swept up below the spout where they had sifted out. An examination of specimens sent showed that three species of predaceous mites were present among the others, and one of them was so numerous that there was no hesitation in writing to the Milwaukee gentleman that the predaceous mites would probably soon destroy the wheat feeders and thus the pest of mites would correct itself. The prediction was speedily verified in part a week or so later, when the correspondent wrote: "As you say, the parasitic mites have largely destroyed the smaller ones, and I suppose when their food is all gone they will die of starvation."

THE CHEESE SKIPPER OR HAM SKIPPER.

(*Piophila casei* Linn.)

A small, glistening, black, two-winged fly lays its eggs on cheese, smoked ham, and chipped beef. The eggs hatch into small white cylindrical maggots which feed upon the cheese or meat and rapidly reach full growth, at which time they are one-fifth of an inch in length. The maggot is commonly called "skipper" from its wonderful leaping powers, which it possesses in common with certain other fly larvæ, all of which are devoid of legs. The leap is made by bringing the two ends of the body together and suddenly releasing them like a spring. In this way it will sometimes jump 3 or 4 inches.

This insect, like so many other household species, is cosmopolitan, and was doubtless originally imported from Europe into the United States.

Careful observations on the life history of this species have been made by several writers. In 1892 Miss M. E. Murtfeldt, whose attention was called to the species on account of the great damage which it was represented to be doing in certain Western packing and curing establishments, studied the life history of the summer generation.[1] The eggs were shown by Miss Murtfeldt to be deposited in more or less compact clusters of from 5 to 15, and also scattered singly. In her observation jars the average number was 30 to a single female, but it is possible that under these abnormal conditions the number was smaller

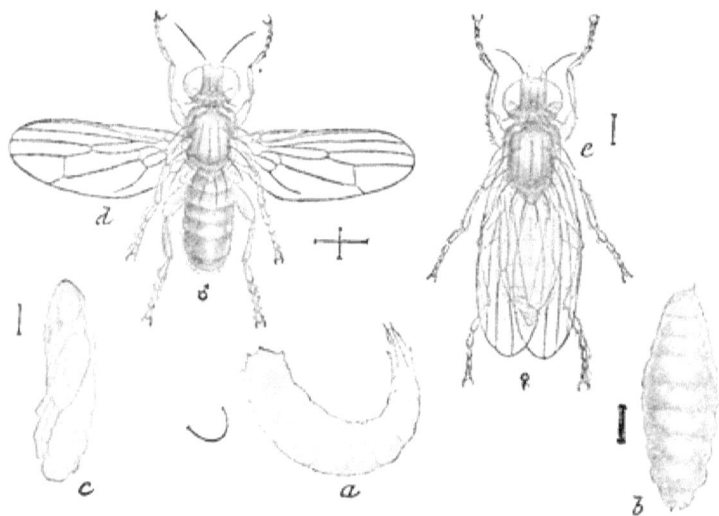

Fig. 48.—*Piophila casei*: a, larva; b, puparium; c, pupa; d, male fly; e, female with wings folded—all enlarged (original).

than usual. The egg is white, slender, oblong, slightly curved, 1 mm. in length, with a diameter of about one-fourth its length. Hatching takes place within thirty-six hours. The larva is cylindrical, tapering gradually toward anterior end, and truncate posteriorly, furnished at hinder extremity with two horny projecting stigmata and a pair of fleshy filaments. The larva completes its growth in from seven to eight days, attaining a length of from 7 to 9 mm. While feeding, if the food supply is sufficient it does not move about much, entire clusters of larvæ often completing their growth in the same crevice in which the mother flies deposited their eggs. When mature, however, it moves away to some dry spot, contracts in length, assumes a yellowish color, and gradually forms into a golden-brown puparium 4 or 5 mm. in length.

In ten days the adult fly issues. Miss Murtfeldt was unable to make the fly lay its eggs on fresh meat of any kind, nor did she find that it was able to breed upon meat which was simply salty. The average duration of adult larvæ, according to her observations, does not exceed a week, and thus the entire life cycle may be concluded in three weeks. These observations were made in August.

During February of the same year specimens of the same insect were sent by a Kansas packing house to Mr. V. L. Kellogg, then of the Kansas State University. At that time of the year his breeding notes show that the egg state occupied about four days, the larva state about two weeks, and the pupa state one week.[1] The adults lived in the breeding jars from six days to two weeks after issuing from the puparia. Larvæ kept with ham and bacon did not take at all kindly to cheese to which they were removed. Careful observations on the life history in Europe have been made by Dr. H. F. Kessler.[2] Dr. Kessler found that the average time in developing from the egg to the adult is four to five weeks, with two or three generations during the summer, the last generation occurring in September, the larva over-wintering in the puparium and transforming to pupa in May. Other writers say that the insect passes the winter in the adult stage.

As a cheese insect in this country this fly does not play as important a rôle as it does as an enemy to smoked meat. It is a matter of observation that the mother fly seems to prefer the older and richer cheeses in which to deposit eggs. Her taste is excellent, and while it is a fair thing to say that "skippery" cheese is usually the best, it will hardly do to support the conclusion that it is good because it is "skippery," although this conclusion is current among a certain class of cheese eaters. With the abundance of the species in packing houses we have nothing to do in this connection. When occurring upon hams it seems to prefer the outer fatty portions.

REMEDIES.

All that we have said of the preventives for the cheese and meat mites will answer equally well for the "skipper." Portions of cheese and hams attacked should be cut out, shelves of pantries should be kept scrupulously clean, and the kerosene-emulsion wash used when it has once been determined that the insect is present in numbers. Every crack should be carefully washed out, since the puparia might be found in such situations. Close screening of the windows of pantries is advised to keep out the fly.

---

[1] Trans. Kans. Acad. Sci., Vol. XIII, 114–115.
[2] Bericht d. Ver. f. Naturk. z. Cassel, Vols. XXIX u. XXX, pp. 58–60.

THE RED-LEGGED HAM BEETLE.

(*Necrobia rufipes* DeG.)

Two or three species of small beetles belonging to the family Cleridæ, and which are normally scavengers, feed occasionally upon dried meats and other stored animal products. The most abundant one in this country is the species indicated in the title. It is a small, rather slender beetle of dark bluish color, with reddish legs. Its larva is a slender worm, and is at first white, with a brown head and two small hooks at the end of the body. As it becomes older it becomes darker, and when full grown is grayish white, with a series of brown patches above. It is then rather more than one-half an inch in length and transforms within a paper-like cocoon. From the appearance of this cocoon the insect has become known as the "paper worm" to dealers in hams and dried meats.

Necrobia rufipes is a cosmopolitan species, occurring all over the United States, in Europe, Australia, Africa, and the East Indies. It is hardly a species that causes a constant drain upon the profits of the trade, but occasionally, under exceptional circumstances, it becomes extremely abundant, and may ruin many hams. It

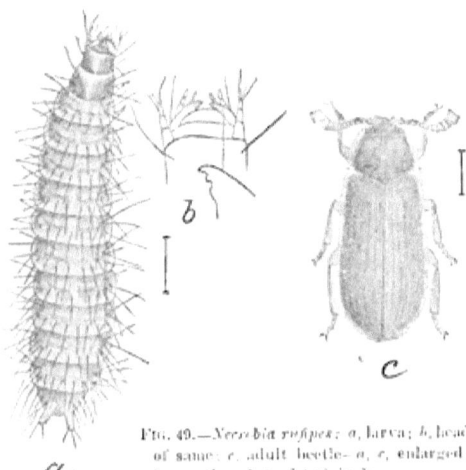

Fig. 49.—*Necrobia rufipes*: *a*, larva; *b*, head of same; *c*, adult beetle—*a*, *c*, enlarged; *b*, greatly enlarged (original).

is by no means uncommon, and is particularly abundant in the West and South.

The injuries caused by this insect are generally due to careless packing of hams or to the accidental cutting or cracking or even to a considerable stretching or fraying of the canvas covering.

As indicated above, this insect is not confined to hams for its food, but lives upon other dead animal matter, not always waiting, however, as do certain other insects, for decomposition to set in before beginning its attacks. The beetle, appearing in May or June, either having bred in the storehouse or storeroom in question, or having flown in from the outside, is attracted to the hams, and wherever it can find the slightest bit of exposed meat it lays a number of minute, narrow, whitish eggs. Such hams as have been injured by overheating or by hanging too long in the sun, from rain, and particularly those which have become slimy from lying too long in the pile, are those which attract it most; but it never seems to lay eggs except where the meat is more or less exposed.

or, at all events, if it does lay the eggs, the young grubs, on hatching, fail to reach the meat, except where they are not obliged to penetrate the canvas.

The larvæ hatch in a few days and burrow into the fatty tissue near the rind, growing rapidly, and seeming to congregate. by preference, in the hollow of the bone at the butt end of the ham. As stated above, they are, when first hatched, white in color, with a brown head and two small hooks at the end of the body. They are slender and very active, and upon reaching full growth they either gnaw into the muscle of the ham or occasionally eat into a neighboring beam, forming a glistening, paper-like cocoon, which appears granulated on the outside. Within this cocoon the larva casts its skin and assumes the pupa state, issuing as a perfect beetle in a longer or a shorter time. According to Dr. Riley, who treated this species in his Sixth Report on the Insects of Missouri (p. 96), there are several generations in the course of a year at St. Louis, but the winter is invariably passed in the larval condition, the first beetles appearing, as previously stated, not earlier than the 1st of May, and usually not before the middle of that month.[1]

REMEDIES

The only remedies which need be insisted upon in case of customary damage to ham by this insect are the early and very careful packing of the hams and the use of strong canvas, impenetrable by the insect, and which is not likely to fray or break. These measures are the direct result of the knowledge of the life history of the insect.

Two instances in the experience of Dr. Riley are of sufficient interest to deserve specific mention. In 1871 and in previous years the firm of Francis Whittaker & Sons, of St. Louis, had suffered serious loss from the damage done by this beetle. After an investigation of the facts they were advised that all of the canvasing on the hams should be done earlier than was customary, or prior to the first of May, and also that a heavier canvas be used, to prevent the possibility of its giving way upon the small ends. This advice was followed, with the result that during the ensuing year not a single ham was lost or returned by a customer on account of worms.

The second case was that of S. S. Pierce, of Boston, who, in May, 1873, received 22 tierces of hams from a Cincinnati firm. The hams were taken from the casks and hung in the loft, and not examined until August, when they were found to be full of worms. Claim was made on the packers for damages, and it was finally agreed to leave the matter to referees, who were selected from prominent packers, and who decided in favor of the Cincinnati firm. The fact is, however, as could

---

[1] Mr. Schwarz states that he has found the adult beetles in the dead of winter in Detroit, Mich., and Cambridge, Mass., and calls our attention to the fact that the species is recorded by H. T. Fay in his article on winter collecting (Proc. Entom. Soc. Phil., Vol. I. p. 197, 1862).

readily have been shown had an expert entomologist been called in, that if the covering of the hams was sound, and had been kept intact while in the hands of the Boston firm, as seems to have been proven by them, the eggs must have been laid before the hams left Cincinnati. The difference in climate between Cincinnati and Boston would also give added weight to the Boston claim. The lack of knowledge of the actual facts governing the case is shown by the written opinion of one of the packing experts, who stated that, whereas the Cincinnati firm had previously used manila paper in packing their hams, they had begun to use husk, which was "very likely to contain the germ from which the worm is bred."!

The insect is hardly a factor in housekeeping except in the country, where a farmer may put up a small number of hams for home consumption during the ensuing year. In ordinary households a wormy ham need only be returned to the dealer from whom it was bought.

THE LARDER BEETLE.

(*Dermestes lardarius* Linn.)

A dark-brown beetle of the shape illustrated in the figure, with a pale, yellowish-brown band containing six black dots across the upper half of the wing covers, three-tenths of an inch in length. The larva is brown and hairy, tapers from head to tail, and is furnished with two short, curved, horny spines on top of the last joint of the body. It is a common museum pest, and is found in many kinds of animal food products, such as hams, bacon, and other kinds of meat, old cheese (of which it seems to be especially fond), horn, hoofs, skins, beeswax, silkworm cocoons, feathers, and hair. It has never been recorded as damaging woolen cloth, and its popular name, "larder" or "bacon" beetle, is a very appropriate one.

The insect has long been known in the United States. It is also found in all parts of Europe and in Asia. It is considered by Dr. Hamilton to be probably a native of the United States as well as introduced by commerce. It seems to occur in all parts of this country.

There are recorded no full and definite statements regarding the life history of this species, and we have made no observations which will enable us to give the length of life, duration of different stages, and other facts of equal interest. Under favorable conditions, however, the insect is unquestionably a rapid breeder. Miss Caroline E. Henstis, of St. John, New Brunswick, in the August (1878) number of the Canadian Entomologist, states that five weeks after placing a female in a glass jar, with a piece of meat, she found a large and flourishing colony of larvæ, most of them full grown. Dr. G. H. Horn, in the Proceedings of the Entomological Society of Philadelphia (Vol. I, 1861, p. 28), states that the insect remains in the pupa condition for a period varying from three or four days to a week, or even more, depending principally on the warmth of the locality. From this statement we see that an entire generation may be developed in six weeks. Therefore, the increase of the insect may be very rapid and there may be four or five generations annually. The larva, when feeding upon dried and smoked meat, according to Dr. Horn, is usually seen creeping on the surface of the meat. For food it prefers such as contains fat and connective tissue, seldom attacking the muscular portions. It does not bury itself in its food until about the time of assuming the pupa state. In general, the beetles make their way into houses in May and June, and at once deposit their eggs on their favorite food if they can obtain access to it. Where this is impossible they will lay their eggs, as will other beetles of the same family, near small cracks, so that the young

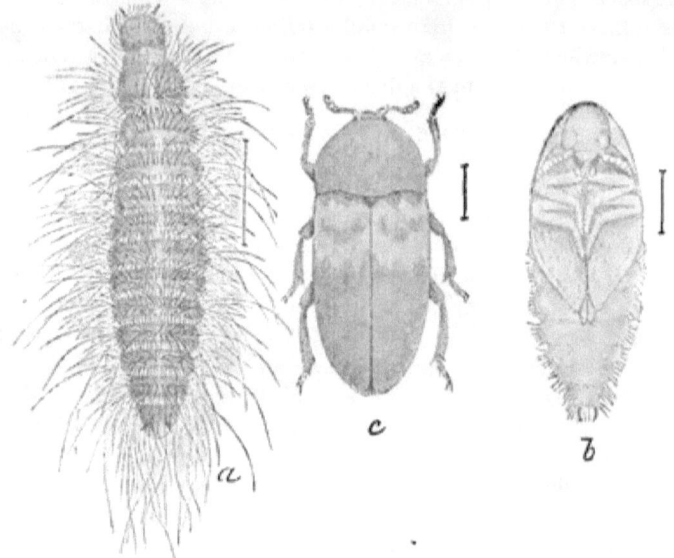

Fig. 50.—*Dermestes lardarius* : *a*, larva; *b*, pupa; *c*, adult beetle—all enlarged (original).

larvæ when hatched can crawl through. Dr. Riley, in his Sixth Missouri Report, states that fresh hams are not so liable to attack by this insect as are those which are tainted or injured.

REMEDIES.

Where a storeroom is overrun with this insect its contents should be cleared out, so far as practicable, and the room should either be sprayed with benzine or subjected to strong fumes of bisulphide of carbon. Where an article of diet such as a ham, has begun to be infested, the affected portion should be cut away and the surface should be washed with a very dilute carbolic solution. Miss Heustis, in the article above-mentioned, showed that tallow was very offensive and destructive to this insect, but there is seldom a case where this interesting bit of knowledge can be utilized. Dr. Hagen, when he first came to Cambridge, found his office overrun with this insect. On a sunny day in

November the southern outer wall was speckled with them. He succeeded in ridding the establishment by trapping them day after day with a piece of cheese. The cheese proved to be extremely attractive, and he destroyed them by hand two or three times a day until he had practically exterminated them. Shortly after the introduction of the Pasteur system of silkworm moth inspection for pébrine in France, according to Maurice Girard, great damage was done by this Dermestid, which attacked first the bodies of the moths as they were attached to their egg receptacles. They laid their eggs in the moths, and their larvæ first ate the bodies and afterwards the silkworm eggs themselves, thus occasioning in 1871 at Pont Gisquet a loss of one-third of the egg crop. The remedial measures adopted were to screen the windows with a very fine wire gauze to prevent the entrance of beetles and afterwards to submit the rooms to fumigation with bisulphide of carbon or corrosive sublimate.

An interesting case of damage to bacon was mentioned by Dr. Lintner in the Cultivator and Country Gentleman for June 26, 1884. An individual in Walkersville, Md., had found bacon hung up in paper meat sacks the 1st of March affected with beetles, and larvæ later in the season, presumably in June. The beetles must have oviposited in the bacon before sacking, or there must have been cracks in the paper bags through which the young larvæ entered. The date of the bagging renders the former hypothesis improbable. The instance seemed to show the necessity for very careful and early bagging. The slightest crack or slit in the paper would be large enough to allow the entrance of the newly hatched larva, since the beetles will lay their eggs near such a crack or slit. Dr. Lintner further advised a thorough whitewashing of the apartment in which the sacks were hung, which in this case was a garret.

THE FRUIT FLIES OR VINEGAR FLIES.

(*Drosophila* spp.)

There are in North America about thirty species of light-brown flies belonging to the genus Drosophila, of which perhaps the majority breed in the juices of decaying and fermenting fruit. Their larvæ are small, white, slender maggots, and are frequently found in canned fruits and pickles which have been imperfectly sealed, occurring mostly near the top of the jars, but living without inconvenience in the briny or vinegary liquid and transforming within brown puparia around the edges of the jar. The commonest species seem to be *D. ampelophila* Loew and *D. amœna* Loew.

The majority of the species are strictly North American, and this includes the two specially mentioned in the paragraph above, although *D. ampelophila* has also been found in Cuba. Several species, however, are common to Europe and the United States, for example, *D. funebris, D. graminum,* and *D. transversa. D. ampelophila* seems the commonest

species all over the United States and is mainly responsible for the
injury to canned fruits and pickles.

All of the species of Drosophila are probably rapid breeders. Careful descriptions of the early stages of *D. ampelophila* and *D. amœna*
are given by Professor Comstock in the Annual Report of the Department of Agriculture for 1881–82. The first-named species he calls the
vine-loving pomace fly, and he met with it frequently in the course of
an investigation of the apple maggot (*Trypeta pomonella*), the flies entering apples which had been injured by the Trypeta, completing the
work of disintegration and hastening decay. They are found commonly, according to Comstock, about the refuse of cider mills and fermenting vats of grape pomace. *D. amœna* he found to be associated
with the former species in apples previously damaged by the Trypeta,
but it was not so abundant as *D. ampelophila*. The larvæ of both

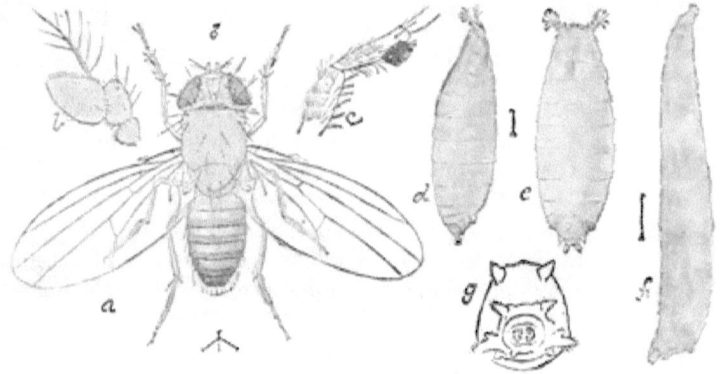

Fig. 51.—*Drosophila ampelophila:* a, adult fly; b, antenna; c, base of tibia and first tarsal joint; d,
puparium, side view; e, same, dorsal view; f, larva; g, anal segment of same—a, d, e, f, much enlarged;
b, c, g, still more enlarged (original).

species, and presumably other species of the genus as well, are furnished with strong anal spiracles through which the larvæ is able to
breathe by protuding simply the end of its body to the air. There are
also delicate tufts about the anal spiracles which may be branchial in
their character.

Professor Forbes, in the Transactions of the Illinois State Horticul-
tural Society, 1884, mentions the damage done by *D. ampelophila* to the
grape crop at Moline, Ill. He states that they attack most frequently
grapes which have been mutilated by birds or damaged by rot, but
once having commenced on a cluster are likely to pass from one berry
to another, the flies meantime constantly laying eggs.

Dr. Lintner, in his first report as State entomologist of New York,
mentions the habits of the European species, showing that *D. cellaris*
occurs in fermented liquids in cellars, such as wine, cider, vinegar, and
beer, and also in decayed potatoes. He also states that a species had
been sent to him as damaging flour paste. He had observed particularly a species which occurred in a jar of mustard pickles. The larvæ,

when nearly full grown, left the liquid and advanced to the side and top of the glass jar where he had placed them, where they could be observed feeding on condensed moisture. They transformed to puparia, from which the first flies issued in four days.

Mr. G. J. Bowles, in the Canadian Entomologist for June, 1882, figures roughly the different stages of *D. ampelophila* and gives an account of its damage to raspberry vinegar. An earthenware jar had been nearly filled with raspberries and vinegar. On opening the jar about ten days later (August 16) it was found to be swarming with the larvæ and cocoons of the insect. Hundreds of the larvæ were crawling on the sides of the jar and the underside of the cover, while pupæ were found abundantly, single and in clusters, particularly where the cover touched the top of the jar. The short time required for the production of so many individuals was surprising. Mr. Bowles half filled a covered tumbler with the pickled raspberries and larvæ, and they continued to produce flies for several weeks. The following season the same observer noticed that the flies were attracted to some raspberry wine in process of fermentation, hovering about the jars and alighting upon the corks, evidently seeking for an opening through which they might pass to lay their eggs. At another time he placed a few raspberries, with a small quantity of vinegar, in a pickle jar with a loose cover. A fortnight afterwards a number of larvæ were seen in the bottle, and several pupæ were attached to its sides.

This statement, together with Dr. Lintner's, that the pupal state may last but four days, shows that a brood may develop in twenty days. The general habits of these insects are well understood by almost every housewife. The writer has often observed them about his own house, and has seen the larvæ working under conditions described by Mr. Bowles, and he is informed by Mr. Marlatt that one of the species is extremely abundant at Manhattan, Kans., and that in his own household the greatest care was necessary to prevent them from entering fruit jars.

REMEDIES.

The common entrance of these little Drosophilas into pantries and storerooms, as well as into dining rooms where fruit is kept upon the sideboard, is another argument in favor of careful window screening. Where they have once entered a jar of fruit it is not necessary to throw away the entire contents of the jar, since the larvæ occur only on the top layers. These may be removed, and the remainder of the contents may often prove pure and sweet. All fruit canned while hot and hermetically sealed will be safe. The flies will lay their eggs upon the jar, perhaps, or upon the cloth covering, and an almost imperceptible opening will suffice for the newly hatched maggot to enter; so the sealing must be perfect. An occasional puffing of pyrethrum about the storeroom will destroy the flies which may have gained entrance. Where a jar has once been opened its contents can be preserved where these insects are numerous only by placing it in some tight receptacle.

CHAPTER VIII.

INSECTS AFFECTING CEREALS AND OTHER DRY VEGETABLE FOODS.

By F. H. Chittenden.

Of the many insects that infest the granary, flouring mill, and warehouse, a considerable proportion contrive at times to find their way into habitations. A small number of these are of almost universal occurrence in the household, and several others are frequently brought into the pantry or storeroom in cereal foods, dried fruits, and other merchandise.

Not so long ago that it has passed out of remembrance it was customary in well-regulated households, even in large cities, to set aside a room, in addition to the cupboard and cellar, for the storage of barrels of flour, bags of meal, boxes of raisins, dried apples, and the like, and such custom still prevails in country homes; but at the present time city housekeepers purchase for the most part in small quantities at the "corner grocery" from time to time as required. As a consequence, the city housewife, unless she should happen to reside in the immediate neighborhood of a store or warehouse, is not so subject to annoyance from storeroom insects as are her country cousins. There is this difference, however, that the farmer's wife is prone to look upon as a necessary evil what the city housekeeper may behold as a veritable calamity. Fortunately, the insects that breed in dry vegetable foods and that display a disposition to make a permanent abode of the storeroom number not more than about a dozen, the remainder, of which a few forms will be selected for passing notice, being only casual visitants and readily controlled under ordinary conditions.

THE FLOUR BEETLES.

Several little flattened beetles of a shining reddish brown color and similar appearance generally so frequently occur in bags and barrels of flour as to have earned the popular title of "flour weevils." They live upon cereal and other seeds and various other stored products, but generally prefer flour and meal and patented articles of diet containing farinaceous matter.

Their eggs are often deposited in the flour in the mills, and these and the larvæ they produce, being minute and pale in color, readily escape notice; but after the flour has been barreled or placed in bags and left unopened for any length of time the adult beetles make their appearance, and in due course the flour is ruined, for when the insects have time to propagate they soon convert the flour into a gray useless mass. A part of the annoyance to purchaser, dealer, and manufacturer is due to the fact that the insects are highly offensive, a few specimens being sufficient to impart a disagreeable and persistent odor to the infested substance.

THE CONFUSED FLOUR BEETLE.

(*Tribolium confusum* Duv.)

The most injurious enemy to prepared cereal foods is undoubtedly the above-mentioned species. Singularly enough, in less than two years from the time of its first recognition as a distinct species occurring in this country, this insect had been reported as injurious in nearly every State and Territory in the Union. The divisional experience of a single year, 1894, shows that more complaints are made of injuries by this than of any other granivorous species. Mr. W. G. Johnson, in the American Miller of January 1, 1896, speaking of this insect as a mill pest, says that it was the most troublesome species of the year 1895, and expresses the belief that it had cost the millers of the United States over $100,000 in manufactured products during that year.

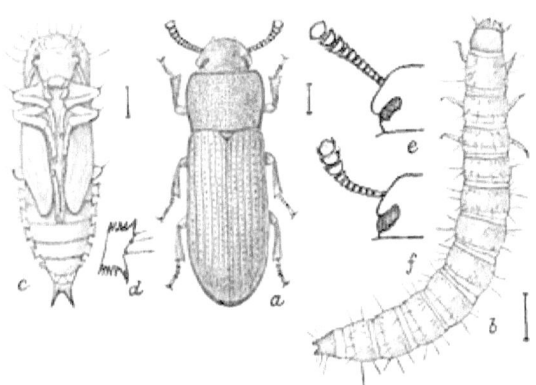

Fig. 52.—*Tribolium confusum*: a, beetle; b, larva; c, pupa—all enlarged; d, lateral lobe of abdomen of pupa; e, head of beetle, showing antenna; f, same of *T. ferrugineum*—all greatly enlarged (author's illustration).

The mature insect is shining reddish brown in color and resembles in miniature the adult of the familiar meal-worm (Tenebrio), which will be referred to further in the following pages. It is scarcely a sixth of an inch long, being almost an exact counterpart of the rust-red flour beetle (*T. ferrugineum*), with which it has been generally confused, but may be distinguished by the structure of the antennæ, which are only gradually clavate, by its broader head, the cheeks being expanded at the sides and angulated at the eyes. The thorax above is gradually narrowed behind, its hind angles being more or less acute. The adult beetle is shown, enlarged, in the accompanying illustration (fig. 52) at a, and the head and antennæ, still more enlarged, at e. The same parts of *ferrugineum* are presented at f for comparison.

This species, like nearly all the others that frequent the family store room, is what is termed a general feeder. It prefers, however, prepared cereals, and hence is most troublesome in flour, corn meal, oatmeal, cracked wheat, and patented foods, but likewise infests in the writer's experience such useful commodities as ginger, cayenne pepper, baking powder, orris root, snuff, slippery elm, peanuts, peas, beans, and seeds of various kinds that are kept long in store. It sometimes also attacks cabinets of dried insects.

As an instance of the nature of injury to flour in households may be mentioned an experience recently reported, as it is one that may fall to the lot of any housekeeper. The house had been closed for six weeks, and on the return of the family the flour, which was kept in a large wooden bucket with tightly fitting cover, and known to the trade as a kanakin, was swarming with the larvæ and beetles of this species. The damaged flour was removed and the bucket refilled, only to be again found with the insects at work in the fresh material. A personal examination showed that the insects, or enough of them to cause reinfestation, had remained in the cracks of the bucket and in holes that they and their larvæ, or *Silvanus surinamensis*, which accompanied them, had made in the soft wood. The bucket was again emptied and the pail scalded, which had the effect of killing all the insects except a few which were discovered to have escaped through holes which they had made in the bottom. The pail had then to be painted on the bottom.

Two reports have reached this office of injury by this species to baking powder. In one instance considerable damage had been done, resulting in the loss of an entire consignment, necessitating its replacement by the manufacturers, not to mention the annoyance to all parties concerned. Customers were returning boxes of the powder almost as soon as opened, on account of the presence of these beetles. The baking powder, of which wheat flour was in this instance one of the ingredients, is put up for sale in tight tin boxes, and so closely covered with paper as to be practically air-tight; consequently the insects must have gained entrance at the manufactory before the boxes were covered.

The life history of this species is in brief as follows: The tiny, clear white eggs are attached to some convenient surface in the cracks or on the sides of the bag, barrel, or other receptacle in which the infested substance is contained. These hatch into minute larvæ, which feed for a period, depending upon the temperature, and then transform to naked, white pupæ, which in due time change to beetles, which copulate soon after transformation, and another generation enters upon its life round. In this manner several broods are generated in the course of a year. From observations conducted by the writer it has been learned that this insect is capable in an exceptionally high temperature of undergoing its entire round of existence from egg to imago in thirty-six days. The minimum period of incubation was not ascertained, but it may be assumed as about six days. This, with six days for the pupal period, gives twenty-four days as the shortest developmental period of the larva. In cooler weather these periods last two or three times as long.

In well-heated buildings in a latitude like that of Washington we thus have the possibility of at least four generations a year.

The mature larva is shown in the figure (fig. 52) at *b*, the pupa at *c* and *d*.

THE RUST-RED FLOUR BEETLE.

(*Tribolium ferrugineum* Fab.)

This species, as previously stated, closely resembles the first-mentioned flour beetle in color, form, and size, but may be distinguished by the form of the head, which is not expanded beyond the eyes at the sides, and by the antennæ, which terminate in a distinct three-jointed club (see fig. 52, *f*). In its habits and life history this insect closely resembles its congener, *T. confusum*, but it is apparently somewhat restricted to the Southern States, although occasionally found in the North. It is often reported in flour, meal, and grain, and is sometimes shipped north in consignments of rice.

THE BROAD-HORNED FLOUR BEETLE.

(*Echocerus cornutus* Fab.)

A third flour beetle that sometimes finds its way into houses is the one above mentioned. It so closely resembles the two preceding species that the females particularly are with difficulty distinguished from them. The male, with its broad, conspicuous mandibular horns, is shown at fig. 53. The general habits of this species also so nearly resemble those of Tribolium that it will be unnecessary to give more than a brief mention of its known foods. It has been found in ground cereals of various sorts, including flour, meal, "germea," rolled barley, bread, army biscuit, maize, wheat, and rice. In southern California it occurs even under bark, showing complete acclimatization. It is somewhat limited in distribution in the United States, but is frequently met with in large seaport towns, especially on the Pacific Coast, and is on the increase elsewhere. In some

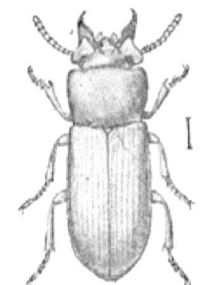

Fig. 53.—*Echocerus cornutus:* male beetle—enlarged (original).

parts of Europe, according to report, it is a veritable pest in bakeries by getting into the flour and into the masses of fermenting dough that accumulate upon the molds used in baking bread.

THE MEAL-WORMS.

Two species of beetles and their larvæ, the latter familiar to nearly everyone under the name "meal-worms," attract attention by reason of their large size and somewhat serpent-like appearance when they invade the family flour barrel, the feed box, bags of bran or meal, or are turned up in unexpected places. These are among the many species that develop in refuse grain dust and mill products that are carelessly permitted to accumulate in the dark corners and out-of-the-way places in flouring mills, bakeries, stores, and stables. The two species are about equally common* and do not differ materially in their habits, and although abundant enough wherever grain is stored, do little or no damage to seed stock, being found mostly in corn meal and other ground products. They are also of some importance as enemies to ship biscuit.

As with some of the other storehouse insects, the Tenebrios are not an unmixed evil, for they have a commercial value to the bird fancier, being used as food for nightingales, mocking birds, and other feathered songsters.

THE YELLOW MEAL-WORM.

(*Tenebrio molitor* Linn.)

The above-mentioned species is the meal-worm most often referred to in scientific literature. Its name—*Tenebrio*, meaning one who shuns the light; *molitor*, a miller—is suggestive of its habits and was given to it by Linnæus in the year 1761 Accounts of its larva, however,

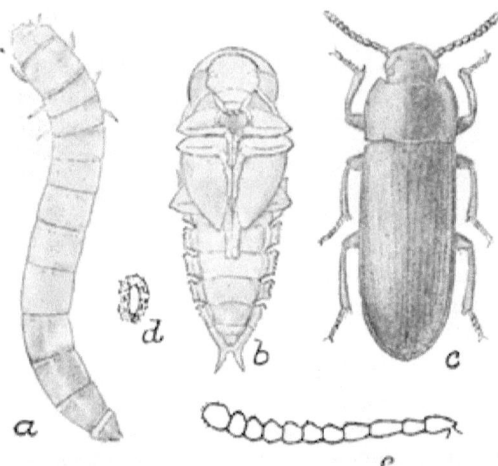

appeared many years earlier, one of these, by Thomas Moufet, dating back to the year 1634. As it is in the larval stage that this insect is best known, the name "yellow meal-worm" is suggested to distinguish it from the congeneric species, which is much darker in color. The larva (see fig. 54, *a*) is cylindrical, long, and slender, attaining a length of upward of an inch and being about eight times as long as broad. It is waxen

Fig. 54.—*Tenebrio molitor: a.* larva; *b,* pupa; *c,* female beetle; *d,* egg, with surrounding case; *e,* antenna—*a, b, c, d,* about twice natural size; *e,* more enlarged (original).

in appearance, much resembling a wireworm. In color it is yellow, shading to darker ochreous toward each end and near the articulation of each joint. The anal extremity terminates in two minute spines, not in a single point, as figured and described by Westwood and other writers. The pupa (*b*) is white, and the adult insect, as will readily be seen by reference to the illustration, (*c*) resembles on a large scale one of the flour beetles. It is considerably over half an inch long, somewhat flattened, shining, and nearly black. An enlarged antenna is shown at *e.*

The eggs are white, bean-shaped, and about one-twentieth of an inch long, and are deposited by the parent beetle in the meal or other substance which is to serve as the food of the future larva, singly or in groups, as high as fourteen or sixteen being laid in a single day. They are adhesive when first extruded and become attached to any surface upon which they are laid, and also take on a coating of particles of meal or other material. In the illustration, at *d*, an egg is shown in profile with its covering of meal.

The beetles begin to appear in the latitude of Washington in April and May, occurring most abundantly in the latter month and in June, when they run and fly actively about in search of their mates and of a new place for the deposition of their eggs. In about two weeks from the time the eggs are laid the infant meal-worm, which is at first clear white in color and with prominent antennæ and legs, makes its appearance. It soon turns yellow, and as it feeds voraciously its growth is rapid. In three months it attains approximate maturity, and from then till the following spring undergoes little change. After having shed about a dozen skins, beginning from soon after its hatching, it changes to pupa and in this state remains about a fortnight. It will, therefore, be noticed that this species is annual in development, a single brood only appearing each year. The beetles are nocturnal, and, being moderately strong flyers, are often attracted to lights. They have the pungent odor characteristic of the family Tenebrionidæ.

In 1889 a physician sent us larval specimens of this meal-worm reported to have been ejected from the stomach of a patient, and there are many other records of similar occurrences of these larvæ in the human body. We also received during the year a specimen of this insect, with an accompanying newspaper clipping giving an account of its having been taken in a hotel from a large pin cushion filled with "shorts." The noise made by the beetles scratching about in endeavoring to obtain their exit from the cushion had caused a guest to complain that his room was haunted. (See Insect Life, Vol. II, p. 148.)

THE DARK MEAL-WORM.

(*Tenebrio obscurus* Linn.)

The darker of the two meal-worm larvæ has been called by writers the American meal-worm, an obvious misnomer, as this species, like the preceding, in all probability came originally from temperate Europe or Asia, and is, like other species most commonly found in the store-house, an introduced cosmopolite.

The mature insect, illustrated at fig. 55, is very similar to the parent of the yellow meal-worm, being of nearly the same dimensions, but distinguishable by its color, which is dull, piceous black. There are other points of difference, notably in the antennæ, the third joint in the present species being perceptibly longer than in *molitor*. The larva also resembles that of the preceding, differing chiefly in its much darker brownish markings. The pupa, however, is of the same whitish color.

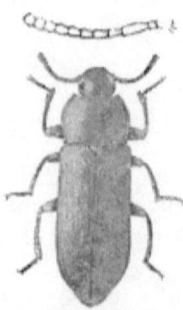

The beetles, in the writer's experience, begin to appear considerably earlier than do those of the yellow meal-worm. Here at Washington they may be found as early as the latter part of February, remaining till the beginning of July, occurring most abundantly in April and May.

In 1890 a correspondent sent specimens of larvæ that had been found in a grocery store in a parcel of adulterated ground black pepper, and within the year we received a lot of living larvæ from Dr. J. B. Porter, of Glendale, Ohio, that had been found in a box of commercial soda ash. We have also specimens that were taken among phosphate fertilizers, cotton seed, and cotton meal. It should be unnecessary to remark that these larvæ did not feed upon the chemicals, although they lived in them for some time.

FIG. 55.—*Tenebrio obscurus:* male—somewhat enlarged (original).

THE MEAL MOTHS.

Two species of moths, in addition to the clothes moths, are habitual frequenters of the household, the one attracting notice through the depredations of its larva in a variety of articles, the other chiefly by its beautiful appearance in the winged form.

THE INDIAN-MEAL MOTH.

(*Plodia interpunctella* Huebn.)

A small moth of about the same size as the clothes moths, which it somewhat resembles when in flight, is very often found in stores, and through them is brought into the household, where it is an all-round nuisance, feeding upon almost anything edible. It makes its home almost everywhere, and is very sure to be found in boxes of preserved fruits if these are left open for any time, but does not disdain fruits that have been left in barrels to rot and dry up, as frequently happens.

The common name of this insect is sufficiently indicative of its fondness for meal, and it feeds as well upon flour and upon grain of all sorts, ground or whole. In the writer's experience it breeds also in chick-peas and table beans, peanuts, English walnuts, almonds, edible acorns, chocolate beans, dried fruits of all kinds, including currants, raisins, peaches, apples, apricots, prunes, plums, and cherries, and seeds of several sorts. It has also been recorded as infesting clover seed, garlic heads, dried roots of dandelion, pecan nuts, and cinnamon bark, and has been reported to invade beehives, and does considerable damage at times in museums, feeding on herbarium specimens, and even attacking dried insects.

The adult moth has a wing expanse of between a half and three-quarters of an inch, and is of the general appearance represented in the illustration (fig. 56, *a*). The outer two-thirds of the fore-wings are reddish brown, with a coppery luster; the inner portion and the hind-

wings are light dirty grayish. The larva, or caterpillar, shown at *c*, *d*, *e*, and *f*, measures when full grown about half an inch and varies in color, being whitish, with light rose, yellowish or greenish tints. The pupa (*b*)is light brown in color.

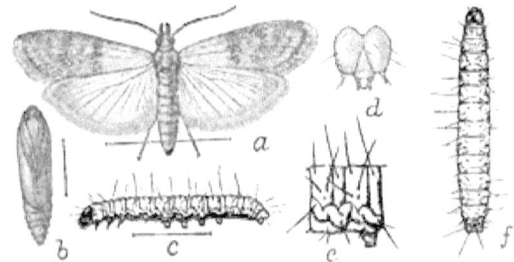

Fig. 56.—*Pbalia interpunctella*: *a*, moth; *b*, chrysalis; *c*, caterpillar, lateral view; *f*, dorsal view—somewhat enlarged; *d*, head, and *e*, first abdominal segment of caterpillar—more enlarged (author's illustration).

The eggs are minute and white, and are deposited, to the number of 350, singly and in groups of from three to a dozen or more, upon whatever substance the female may see fit to select for the sustenance of her offspring. In four or more days they hatch, and in four or more weeks another brood is produced. In this manner a succession of generations appears which will vary, according to the temperature of the building that the insect inhabits, from four to possibly six or seven a year.

The caterpillars spin a certain amount of silk as they feed, joining together particles of their food and excrement, and thus injure for food several times the amount of material that they consume. When fully matured they crawl hither and thither, trailing large quantities of their silken threads after them, in their search for a suitable place for transformation, and finally surround themselves in a cylindrical silken web, in which they change to chrysalids and then to moths.

THE MEAL SNOUT-MOTH.

(*Pyralis farinalis* Linn.)

This species in its mature condition is the most attractive of all household insects. It measures across its expanded fore-wings upward

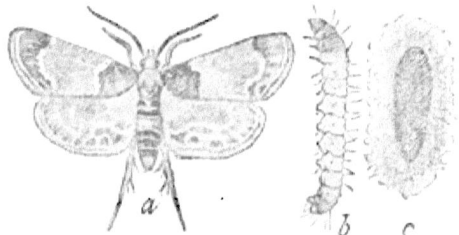

Fig. 57.—*Pyralis farinalis*: *a*, adult moth; *b*, larva; *c*, pupa in cocoon—twice natural size (original).

of three-quarters of an inch. Its dark colors are of different shades of brown, with reddish reflections; the lighter colors are whitish and form the pattern shown in the illustration (fig. 57, *a*). The caterpillar (*b*) is whitish, shading off to a darker color at either end, and with a reddish head. The pupa, shown in its enveloping cover of silk at *c*, and naked at fig. 58, *e*, is reddish brown.

The habits of this moth are peculiar. The larvæ subsist chiefly upon cereals, but seem not to prefer them in any particular condition, feeding alike on the seed, whole or ground, bran, husk, or straw. They will attack other seeds and dried plants, and are at times injurious to hay, particularly clover. They are also reported to feed upon stored potatoes. Within the year larvæ were brought to this office in flour and specimens of the insect's work in sweet marjoram, an herb sometimes used in cooking. The caterpillars live in long tubes or tunnels composed of silk and particles of meal or other material, and while thus incased in the obscure corners in which they habitually live are completely concealed from observation. When mature they leave them and construct cocoon-like cases and undergo transformation within.

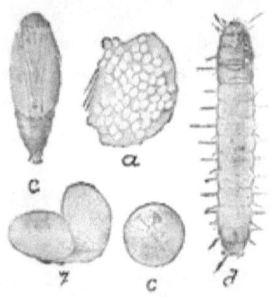

FIG. 58.—*Pyralis farinalis: a, egg-mass; b, eggs, more enlarged; c, egg showing embryo within; d, larva, dorsal view; e, pupa—all enlarged (original).*

The life history of the meal snout-moth has never been properly understood, the efforts to rear and observe it having always proved unsatisfactory. Certain European writers have expressed the belief that the species is biennial in development, but experiments now being conducted go to prove at least four generations a year. The species has been carried through all its stages this spring in about eight weeks.

From recent experience it would seem that comparatively little danger need be apprehended from injuries by this insect if material upon which it is likely to feed be kept in a clean, dry place. Almost without exception, the cases of damage attributable to it have occurred in cellars, upon floors, in outhouses, or in places where refuse vegetable matter had accumulated.

THE GRAIN BEETLES.

There are two clavicorn beetles, known, respectively, as the saw-toothed grain beetle and the cadelle, of omnivorous habits and universal distribution, that commonly occur in dwellings as well as in granaries, mills, and warehouses. The former is so small as to readily escape notice except when it is present in numbers; the latter, though seldom occurring in abundance, is conspicuous, both as larva and beetle, on account of its size. The two species resemble each other in being partially carnivorous and predaceous, following in the wake of other insects like the Indian-meal moth, the cadelle particularly making atonement for its ravages in the pantry supplies by devouring such small insects as cross its path that it is able to overcome.

THE SAW-TOOTHED GRAIN BEETLE.

(*Silvanus surinamensis* Linn.)

Taken all in all, this is perhaps the commonest insect that habitually abides in groceries, and, excepting the so-called Croton bug, the one most often found in the pantry. Wherever anything edible is stored this insect will be found. It is chiefly vegetarian, but is almost omnivorous, and is especially fond of cereals and breadstuffs, preserved fruits, nuts, and seeds of various kinds. Among other commodities of the household that are subject to its depredations may be mentioned yeast cakes, mace, snuff, and red pepper.

The mature beetles will feed upon sugar and have been reported in starch, tobacco, and dried meats, but it is doubtful if the insect will breed in such substances. The beetles or their larvæ have the bad

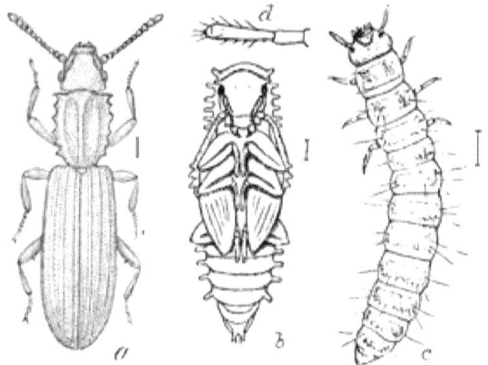

Fig. 50.—*Silvanus surinamensis: a*, beetle; *b*, pupa; *c*, larva—all enlarged; *d*, antenna of larva—more enlarged (author's illustration).

habit of perforating the paper bags in which flour and other comestibles are kept. When present in boxes of fruit—and they are very sure to be there if the covers are left off thoughout the summer—there may be no visible evidence of their presence until the bottom is reached, but here they will be found in great numbers, and when disturbed scamper off in the greatest haste. This insect is almost invariably present wherever the Indian meal moth is found, and the list of the food products that have been mentioned as subject to this moth's attack will answer about equally well for the beetle.

As an instance of unusual trouble caused by this insect may be mentioned the case cited by Taschenberg of the beetles having invaded sleeping apartments adjoining a brewery where stores were kept and annoying the sleepers at night by nipping them in their beds.

This beetle is a member of the family Cucujidæ. It is only about one-tenth of an inch long, slender, much flattened, and of a chocolate-brown color. The antennæ are clavate, or club-shaped, and the thorax has two shallow, longitudinal grooves on the upper surface and bears six minute teeth like those of a saw on each side, as indicated at fig. 59, a.

The larva is somewhat depressed, and nearly white in color, with darker markings, as shown in the illustration (c). It has six legs and an abdominal proleg, and is exceedingly active, running about, nibbling here and there.

When fully matured the larva fastens itself by means of some adhesive matter, evidently excrementitious, to any convenient surface, and thus attached transforms to pupa and afterward to imago. When the insect is living in such granular substances as oatmeal and cracked wheat a delicate case is constructed of fragments of these materials, but when in flour and meal often no covering is made. From data acquired by experiment it is estimated that there may be six or seven generations of this insect annually in the latitude of the District of Columbia. During the summer months the life cycle requires but twenty-four days; in spring, from six to ten weeks. At Washington, it has been learned, the species winters over in the adult state, even in a well-warmed indoor temperature.

THE CADELLE.

(*Tenebroides mauritanicus* Linn.)

The term "cadelle" was first proposed years ago in France for the larva of this insect. The Latin name was given to it in 1758, when it was described as a species of Tenebrio and classified with the mealworms, the adult of which it very slightly resembles in its somber color and depressed elongate form. It belongs, however, to a distinct family, the Trogositidæ, and is considerably smaller than the mealworm beetles, measuring about a third of an inch. It is very dark, shining brown in color, much flattened, and of the somewhat oblong form indicated in the illustration (fig. 60, a). The antenna is shown, much enlarged on page 123. The general appearance of the larva is shown at c. It is fleshy and slender, measuring when full grown nearly three fourths of an inch. It is whitish in color, with head and tip of the anal segment dark brown, the latter terminating in two dark corneous hooks. The three thoracic segments are also marked with dark brown, as indicated in the figure. The pupa (b) is white.

There has always been a difference of opinion in regard to the nature of the food of *Tenebroides mauritanicus*, some claiming that the insect was carnivorous. It has been satisfactorily proven through experiment by the writer that the insect is both herbivorous and predaceous. It is most often found in cereals and in nuts, but may be occasionally taken in other materials.

If personal experience and divisional records be any criterion, this species excels all other grain feeders in its proclivity for obtruding its presence in unexpected places. It is a most unwelcome guest at all times, its large size, both in the larval and adult stages, rendering its appearance conspicuous, not to say alarming or disgusting, to most persons. In the pages of Insect Life we have noted its presence in milk (Vol. I, p. 112), the evidence being that the milk had been adulterated with some farinaceous material in which the beetle had lived as larva. On pages 314 and 360 it is mentioned as having tunneled for a long time through a flask of an insecticide (white hellebore) which was

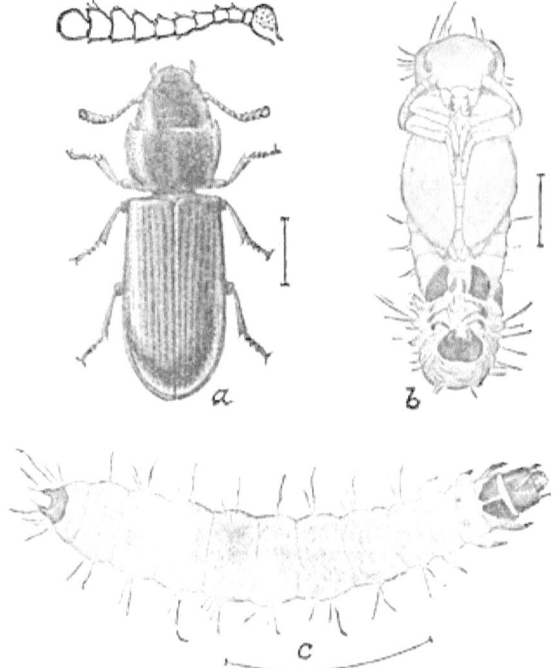

Fig. 60.—*Tenebroides mauritanicus*: a, adult beetle with greatly enlarged antenna above; b, pupa; c, larva—all enlarged (original).

found by experiment to be of sufficient strength to kill currant worms. Again, on pages 274–275 of Volume VI we note the presence of this and other insects in refined sugar. Mr. R. S. Clifton, of this office, recently showed the writer a larva found in powdered sugar, with the information that the sugar had been returned promptly to the grocer of whom it had just been purchased. In granulated sugar the occurrence of this and probably of other insects is generally the result of accident, as it has never been proven that insects breed in sugar in this condition. In the case of pulverized sugar, however, the presence of insects would at least create a suspicion of adulteration with flour.

Contrary to the rule with regard to indoor species, there is every reason to believe that this insect is of American nativity. It differs also from most other storehouse species in being annual in its development, propagating, it is true, throughout the warm season, but bringing forth only a single brood each year.

THE DRUG-STORE BEETLE AND ITS ALLIES.

THE DRUG-STORE BEETLE.

(*Sitodrepa panicea* Linn.)

One of the commonest of storehouse pests is the little *Sitodrepa panicea*, a frequent visitor in habitations, which it enters at open windows.

This beetle is a member of the family Ptinidæ. It is cylindrical in form, measuring about a tenth of an inch in length, and is of a uniform light-brown color, with very fine, silky pubescence. The elytra are distinctly striated and the antennæ terminate in an elongate three jointed

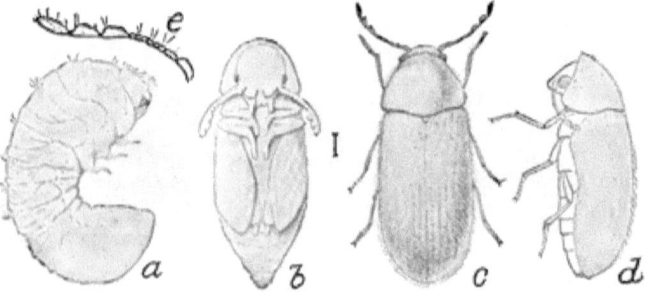

Fig. 61.—*Sitodrepa panicea:* a, larva; b, pupa; c, beetle, dorsal view; d, lateral view—all much enlarged; e, antenna—more enlarged (original).

club. Fig. 61, *c*, shows the beetle with antennæ extended, *e* representing an antenna greatly enlarged. When at rest the head is retracted into the peculiar hood-like thorax, as shown in profile at *d*, and with the legs and antennæ folded under and tightly appressed to the body, the little creature easily escapes observation. The larva is white, with darker mouth-parts, and of the cylindrical curved form indicated at *a*. The characteristic form of the head and legs is reproduced at fig. 62. The pupa, illustrated at *b*, is white.

The insect received its Latin name from its occurrence in dry bread (*panis*), and in Europe it is still known as the bread beetle, but its chief injuries are to druggists' supplies; hence the name drug-store beetle. Its depredations do not stop here, however, for it invades alike stores of all kinds, mills, granaries, and tobacco warehouses. Of household wares its preference is for flour, meal, breakfast foods, and condiments. It is especially partial to red pepper, and is often found in ginger, rhubarb, chamomile, boneset, and other roots and herbs that were kept in the farmhouse in our grandmothers' day. It also sometimes gets into dried beans and peas, chocolate, black pepper, powdered coffee, licorice, peppermint, almonds, and seeds of every description.

The subject of injuries wrought by this species has formed the text of a considerable literature, going back to the year 1721, when Pastor Frisch found the larva feeding upon rye bread, and including, besides damage of the nature referred to, injury to drawings and paintings, manuscripts and books. Some singular instances are recorded of its injuries as a bookworm. The late Dr. Hagen wrote that he once saw "a whole shelf of theological books, two hundred years old, traveled through transversely" by the larva of this insect, and still another record is published of injury by this species, or *Ptinus fur*, to twenty-seven folio volumes, which it is said were "perforated in a straight line by one and the same insect, and so regular was the tunnel that a string could be passed through the whole length of it and the entire set of books lifted up at once."

In pharmacies it runs nearly the whole gamut of everything kept in store, from insipid gluten wafers to such acrid substances as wormwood, from the aromatic cardamom and anise to the deadly aconite and belladonna. It is particularly abundant in roots, such as orris and flag, and sometimes infests cantharides.

It is recorded to have established a colony in a human skeleton which had been dried with the ligaments left on, and the writer has seen speci- mens taken from a mummy. It has even been said to perforate tin foil and sheet lead, and that it will "eat anything except cast iron." In short,

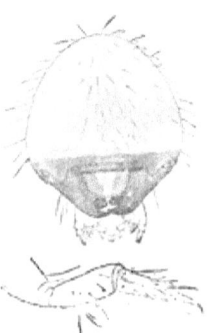

Fig. 62.—*Sitodrepa panicea*; Head of larva, shown above; leg of larva below—much enlarged (original).

a whole chapter could be devoted to the food material of this insect, as nothing seems to come amiss to it and its voracious larva. The sub- ject may conclude with the statement that this Division has received complaints from three different correspondents of injury to gun wad- ding, and there are several records of injury to boots and shoes and sheet cork.

The larvæ bore into hard substances like roots, tunneling them in every direction, and feed also upon the powder which soon forms and is cast out of their burrows. In powdery substances the larvæ form little round balls or cells, which become cocoons, in which they undergo transformation to pupæ and then to the adult insect. I have reared the insect from egg to beetle in two months, and as it habitually lives in artificially heated buildings and breeds out through the winter months, there may be at least four broods in a moderately warm atmosphere.

Minute as is this beetle, it is preyed upon by a still smaller parasite, a chalcis fly known as *Meraporus calandra* How., which pursues its victim relentlessly, even entering insect boxes infested by its host, as the writer had once occasion to observe. A diminutive mite, *Heteropus ventricosus* Newp., also preys upon this as well as many other species of like habits, attacking it in its larval and pupal condition.

THE CIGARETTE BEETLE.

(*Lasioderma serricorne* Fab.)

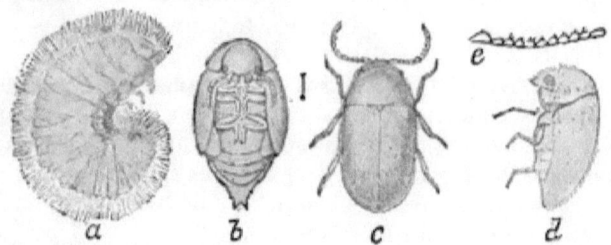

FIG. 63.—*Lasioderma serricorne:* a, larva; b, pupa; c, beetle; d, same, lateral view—all enlarged; e, antenna—much enlarged (original).

Another little beetle, superficially resembling the preceding species and having very similar habits, often occurs in houses. As its English name indicates, it is chiefly known as a destroyer of tobacco, and as such, in the opinion of many thinking people, should be classified with beneficial insects. It is by no means so common as the drug-store beetle, but it is on the increase and doubtless will in time be found to have nearly the same range of food materials. As a tobacco feeder it outranks that species, and also appears to favor certain medicinal plants not so often affected by the Sitodrepa.

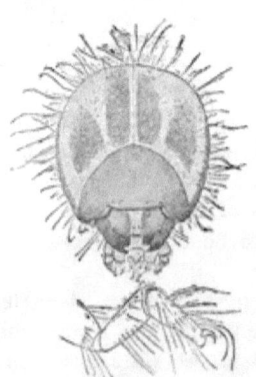

FIG. 64.—*Lasioderma serricorne:* Head of larva, shown above; leg of larva below—much enlarged (original).

Of household supplies it has been found, in the experience of the writer as well as of others, infesting cayenne pepper, ginger, rhubarb, rice, figs, yeast cakes, and prepared fish food. It has been reported as destructive to silk and plush upholstery, and the past year did considerable damage to dried and preserved herbarium specimens in Washington. Of drugs it is partial to ergot and turmeric, and tobacco it devours in every form, in the leaf and when made up into chewing plug, cigarettes, and cigars.

This species is of about the same size and color as the drug-store beetle, but, as may be seen in the figure (63, c), is more robust and the elytra are not striated. The head is more prominent and the antennæ are nearly uniformly serrate, not ending in a three-jointed club (fig. 63, e). The larva, represented at d in curved position at rest, is more wrinkled and hairy than that of Sitodrepa, and differs as well in the structure of the head and legs (see fig. 64). The pupa, shown at fig. 63, *b*, is white and is incased, like other ptinids, in a fragile cocoon.

THE WHITE-MARKED SPIDER BEETLE.

(*Ptinus fur* Linn.)

Two more species of this same family and of somewhat similar habits to the two beetles just mentioned are sufficiently common in storerooms and cellars, particularly of old houses, and especially in the North, to attract occasional notice. The more important of these is *Ptinus fur*, which may be called the white-marked spider beetle, to distinguish it from the allied *Pt. brunneus*, which is uniform brown in color. This beetle is reddish brown, with four white bands on its elytra. It has long antennæ and legs and a more or less globular body, and strongly suggests a spider in general appearance. The sexes differ considerably, the female being much more robust than her consort.

As long ago as 1766 Linnæus gave an account of this species, which he stated was very injurious in libraries. It occurs also in old barns, warehouses, and museums, and is credited with feeding upon a variety of substances, vegetable and animal, including insect collections and dried plants in herbaria. It has also been recorded as living in boxes of red pepper, and during March of the present year was so reported by Mr. R. C. Lyle, who furnished us with specimens in the infested substance brought from his home at Cedar Springs, Mich. Many years ago it was severely injurious to flour at Versailles, France, and two years since Mr. James Fletcher received complaints of its occurring abundantly in flour at Orillia and Toronto, Canada.

During 1894 we received specimens of this insect, with information that they had been discovered near Concord, N. H., in a barn in which were stored a hundred or more bags of cotton seed. They had devoured the bags and increased so enormously as to cover the buildings; had invaded neighboring houses, and were attacking clothing of all kinds. The owner of this barn, who also conducted a store, was greatly alarmed for fear they would spread throughout the town, and serious apprehension was felt in the infested locality that the insect might become a public nuisance.

When to the items just mentioned we add that Dr. George Dimmock found this species swarming in a barrel of refuse wool covered with sheep's dung, and in which it was doubtless breeding, and that, to the writer's personal knowledge, the adults are attracted to fresh fruit, we sum up the principal facts known regarding this insect in America; but if we are to believe all the bad things that are said of it in Europe, it is capable of becoming a serious pest if once permitted to gain sufficient headway, for it is accused of depredating upon furs and clothing, roots, grain, and stuffed animals, and of invading seed stores, apothecaries' wares, and cracker stores.

The larva is white and of the usual ptinid form, quite similar to that of the drug-store beetle, and feeds, like that species, in a little globular case of delicate construction and composed of the material that it infests, and which it cements loosely together. The development of this species is said to be annual in Europe. It has been carried through all its transformations here at Washington in about three and a half months, the pupal period lasting thirteen days.

The adult beetles are nocturnal and may be found in the dead of winter crawling upon the walls of cellars and unheated buildings.

THE BROWN SPIDER BEETLE.

(*Ptinus brunneus* Duft.)

The last of the domestic Ptinidæ that will require special notice is the one above mentioned, and which, as previously stated, differs from its congener chiefly in lacking the white marking on its elytra. Nor is there probably any degree of difference in habits and life history beyond the recorded list of food materials observed for each species. Both occur in the same locations, not unusually living together in apparent harmony. Like *Pt. fur*, it is disposed to be omnivorous and is somewhat of a scavenger, frequenting cellars and attics, storehouses, henhouses, and pigeon lofts, being competent to eke out a living almost anywhere where anything animal or vegetable is stored. Among the different substances that afford it sustenance are books, feathers, skins, dried mushrooms, and the excrement of rats and other domestic animals. It sometimes gets into drugs, and is recorded to have attacked musk root and the powdered leaves of senna and jaborandi.

SPECIES OF OCCASIONAL OCCURRENCE IN VEGETABLE STORES.

The following insects are so often found in dry vegetable foods as to deserve brief mention. Like preceding species, they are cosmopolitan in distribution and occur in the greatest numbers in tropical climates.

The granary weevil (*Calandra granaria* Linn.), a small dark-brown species about an eighth of an inch long, is very partial to the pearled barley used in the preparation of soups, and the chick-pea, a leguminous seed cultivated for the same purpose in tropical countries.

A similar species, the rice weevil (*C. oryzæ* Linn.), which, with the preceding, is most destructive in stored grain, as an adult insect sometimes invades boxes of cakes, crackers, yeast cakes, macaroni, and similar breadstuffs, and is said to attack chestnuts, bird seed, and even to injure tobacco. It also breeds in rice and in cracked corn and other cereals that are sufficiently coarse for the purpose.

Two weevils belonging to the family Bruchidæ, of wide distribution, and known respectively as the pea weevil (*Bruchus pisorum* Linn.) and the bean weevil (*B. obtectus* Say), lay their eggs upon ripening peas and beans in our gardens and thence find their way to our tables, being often eaten when in the larval condition, safely screened from view in these esculent legumes. The former species is restricted to the pea for food, and though it passes the winter in peas that are kept in store, does not breed, as does the latter, for successive generations in the same seed.

Still another weevil (*Araecerus fasciculatus* DeG.), a member of the family Anthribidæ, and for which is proposed the name "coffee-bean weevil," occurred in abundance during the past year in a local grocery store, having been reported to us by a purchaser who found numbers of the beetles in dried apples. This species infests, besides coffee beans and dried apples, mace, nutmegs, chocolate beans, and the roots of a species of ginger.

Certain species of Dermestidæ, it has recently been learned, in addition to a diet of dried animal matter, attack cereals and other vegetable products. The commonest of these is the black carpet beetle (*Attagenus piceus* Oliv.), an account of which, by Dr. Howard, has appeared in preceding pages. Its larva breeds in cereals, ground and whole, and has been reared from millet, pumpkin, and timothy seed. *Trogoderma tarsale* Melsh. has similar habits, and has been found living in grain, flaxseed, castor beans, cayenne pepper, millet and pumpkin seeds, peanuts, and meal and cake manufactured from them. *Anthrenus verbasci* Linn., a near relative of the so-called "buffalo moth" treated in previous pages, has nearly the same food habits as the two preceding species.

A grain beetle known as *Cathartus advena* Waltl, of the same family as *Silvanus surinamensis*, has similar habits to the latter, but is much rarer in stored products. It has been taken by the writer in dry dates, figs, and cacao beans.

Læmophlæus pusillus Sch., another cucujid beetle, smaller, flatter, and with longer antennæ than the preceding, occurs in flour, meal, grain, etc., but, as it is at least partially predaceous, does little harm.

Several small species of the family Nitidulidæ are at times very injurious to dried fruits, but seldom occur abundantly in this country, except in the South. One of the commonest of these is *Carpophilus hemipterus* Linn.

A gray moth of the genus Ephestia, related to the Indian-meal moth, sometimes occurs with this latter in nuts and fruits. It is about equally common in English walnuts, and its pinkish-striped larvæ do considerable injury to dried figs.

The Angoumois grain moth (*Sitotroga cerealella* Ol.), a destructive granary insect, is very injurious to popcorn, and infests also rice and and other cereals.

REMEDIES.

A considerable percentage of injury to the dried vegetable products of the household may be prevented by a moderate degree of care when purchasing, and in storing in tight receptacles in cool, dry rooms.

2805—No. 4——9

The vegetable foods most subject to injury are prepared cereals. If any of these be badly infested at the time of purchase it will be plainly evident: if only a moderate number of insects be present and it be desirable to store the material for some length of time, by sifting over a large sheet of paper of light color, using a fine sieve for flour and corn meal and a coarser one for cracked wheat and like foods, the presence of infesting insects may be detected.

After what has been said regarding the development of the flour beetles and other insects it should be superfluous to add that it is impossible to entirely free infested material by sifting, as the eggs and younger larvæ slip through the finest meshes. Most insects may be destroyed by placing the material infested in the oven at a moderate degree of heat, from 125° to 150° F., but care must be exercised not to expose it to a higher temperature. Corn meal, particularly, is easily overheated, and afterwards, unless it is soon to be used in cooking, is apt to become rancid.

If a barrel of flour or large quantity of other provisions becomes infested, as is apt to happen during the absence of a family from home, bisulphide of carbon, a liquid chemical in general use against insects in mills, elevators, granaries, and warehouses, should be used to disinfect it. The same reagent is the best insecticide for use when whole rooms are to be fumigated.

A small quantity of the chemical is sufficient for the disinfection of a barrel of flour, as the insects for the most part live only in the flour at the top, being unable to withstand the pressure of a large weight of material. From a half to a whole teacupful (about 2 to 5 ounces) of the bisulphide will prove sufficient for the purpose in an ordinary case, provided the cover be replaced as tightly as possible. In more severe cases of infestation it may be necessary to repeat the application. The bisulphide is poured into shallow pans or plates placed upon the top of the infested mass and the receptacle covered as closely as possible and left for a day or more. This chemical is extremely volatile, and being heavier than air, descends as a gas, killing such insects as the material may contain. When an entire room or building is overrun with insects, the bisulphide is evaporated at the rate of a pound to every 1,000 feet of cubic space.

The vapor of this chemical is deadly to all animal life, but there is no danger in inhaling a small quantity, and although it has a powerful and disagreeable odor, this soon passes away without any after effects and without harming for food such material as it may be used upon. The vapor is also inflammable, but if no fire, as. for example, a lighted cigar, be brought into the immediate vicinity until the fumes have entirely disappeared, no trouble will be experienced.

Bisulphide of carbon costs, at retail, from 20 to 30 cents a pound; at wholesale, in 50-pound cans, 10 cents a pound.